Introduction

The F-14 Tomcat was *the* American jet fighter during the 1980s. Those big intakes, the hawkish nose, and those wild variable geometry wings – it was the future of air combat plain and simple. And then there was Tony Scott's iconic film Top Gun starring Tom Cruise, Kelly McGillis and Val Kilmer; a whole generation grew up to the sound of Kenny Loggins urging whoever would listen to 'ride into the danger zone'.

That was my impression of the US Navy growing up – the skill and bravado of the aviators, certainly, but the F-14 was the pinnacle of naval fighter development.

I knew that before the Tomcat there was the unforgettable F-4 Phantom II. This was a real dog of war aircraft, bloodied in Vietnam and brutally powerful. But it had become ubiquitous by the time I became aware of it – the RAF and Royal Navy in Britain both operated it along with numerous other countries all around the world. It somehow lacked the high-tech appeal of the F-14.

Later there came the F/A-18 which, if my Airfix models were any basis for comparison, had a somewhat weedier airframe and whose star turn on the big screen came courtesy of the horrendous Independence Day. It just wasn't in the same league as the Tomcat.

And that was it. For some reason my mind labelled everything prior to the F-4 'old stuff' and I paid it no mind until, one day, I found myself working on the *Aviation Classics* series of magazines – writing material for editor Tim Callaway as well as editing his work. Tim was very good at recapping the full histories of the aircraft he was writing [illegible] and the F-4 issue was a real eye-opener.

The history of McDonnell and later McDonnell Douglas was fascinating – it turns out there was a whole crucial chapter of aviation history I had foolishly been ignoring for years!

Looking back from 2020, a time when the lifespan of modern fighter aircraft is measured in decades, if not half-centuries, it seems strange to think that once upon a time a US Navy jet fighter was expected to last no more than five or six years, a decade tops.

Right from the early days of jet propulsion, the Navy was pushing the boundaries of what was technologically possible with designs such as the Cutlass (not quite as 'gutless' as you might have been led to believe), Demon (ancestor of the F-4) and Skyray (in front line service for just six years!). Clearly, here were aircraft which foreshadowed the mighty Tomcat itself when it came to high-tech prowess.

Today's Navy is finally emerging from what has been a time of unprecedented austerity with the F/A-18E/F and F-35C at the cutting edge of aviation technology and with the F/A-XX programme set to deliver a sixth generation fighter that will take the Navy well into the 21st century.

This publication chronicles the US Navy's jet fighters through the beautiful artworks of renowned aviation illustrator JP Vieira. I hope you enjoy marvelling at the incredible variety of designs as much as I have.

Dan Sharp

ABOUT THE ARTIST

JP Vieira is an illustrator producing military history and aviation-themed artwork.

He is entirely self-taught and aims to constantly improve both the technical and digital methods. His attention to detail and constant pursuit of improvement makes his artworks both accurate and artistically pleasing.

JP is a published artist, collaborating with several authors, editors and publishers.

BOEING F/A-18E/F SUPER HORNET

A nest of Hornets – US Navy air power on display aboard the USS *Abraham Lincoln* on the Arabian Sea in August 2019.
US Navy photo by Mass Communication Specialist 3rd Class Dan Snow

CONTENTS

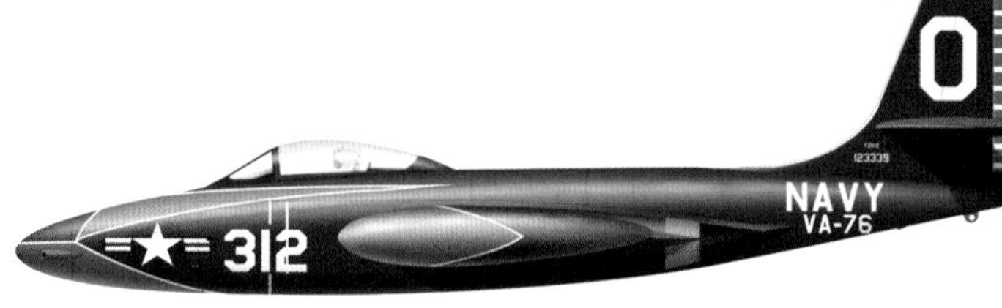

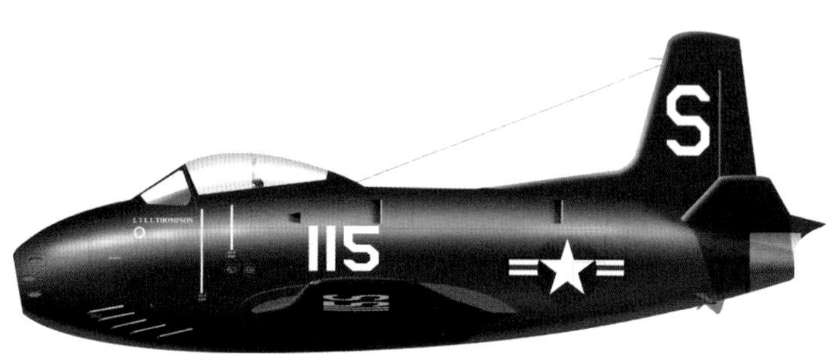

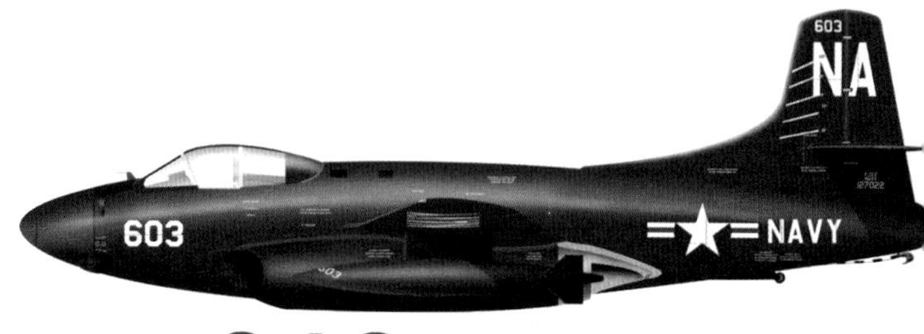

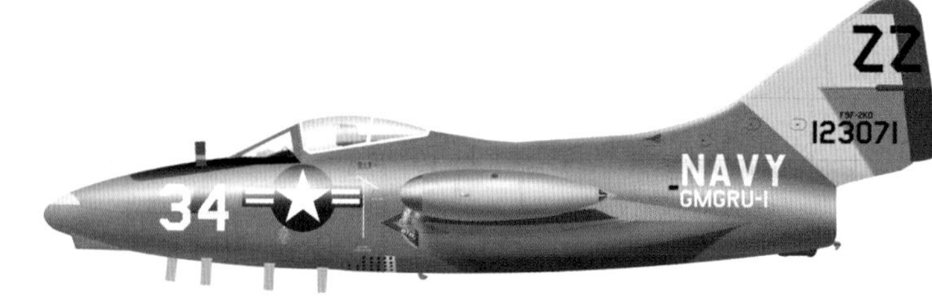

All illustrations:
JP VIEIRA

Design:
BOOKEMPRESS

Publishing director:
DAN SAVAGE

Publisher:
STEVE O'HARA

Picture desk:
JONATHAN SCHOFIELD, PAUL FINCHAM

Production editor:
DAN SHARP

Marketing manager:
CHARLOTTE PARK

Commercial director:
NIGEL HOLE

Published by:
MORTONS MEDIA GROUP LTD, MEDIA CENTRE, MORTON WAY, HORNCASTLE, LINCOLNSHIRE LN9 6JR.

Tel. **01507 529529**

Printed by:
WILLIAM GIBBONS AND SONS, WOLVERHAMPTON

MORTONS MEDIA GROUP LTD

ISBN: 978-1-911639-34-3

© 2020 MORTONS MEDIA GROUP LTD. ALL RIGHTS RESERVED. NO PART OF THIS PUBLICATION MAY BE REPRODUCED OR TRANSMITTED IN ANY FORM OR BY ANY MEANS, ELECTRONIC OR MECHANICAL, INCLUDING PHOTOCOPYING, RECORDING, OR ANY INFORMATION STORAGE RETRIEVAL SYSTEM WITHOUT PRIOR PERMISSION IN WRITING FROM THE PUBLISHER.

NPF
142448
NAVY

XF
NAVY
VX-4
23
129704

NATF
3575
NAVY
4

CHINA
LAKE
0747
NAVY
747

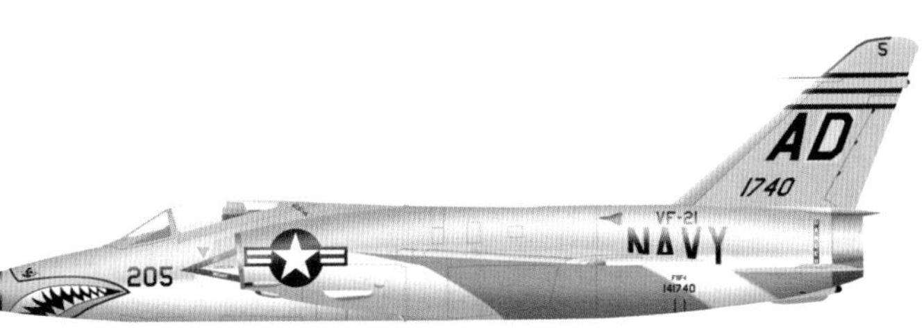
AD
1740
VF-21
NAVY
205
141740

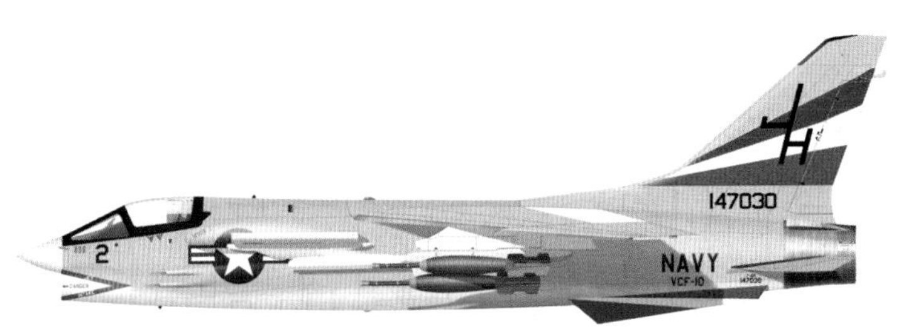
147030
NAVY
VCF-10
2

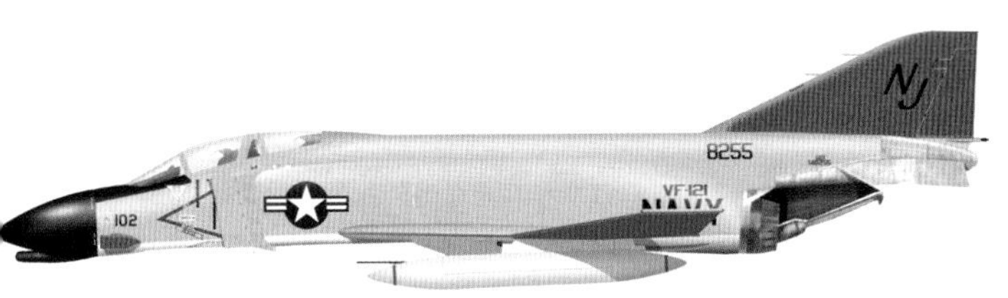
NJ
8255
VF-121
102

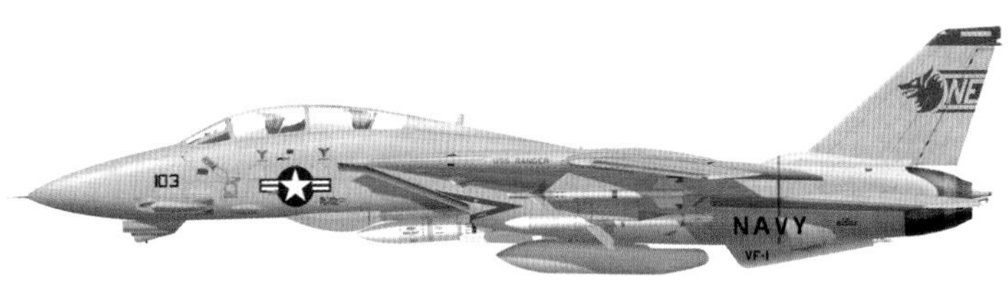
103
NAVY
VF-1

01
NAVY

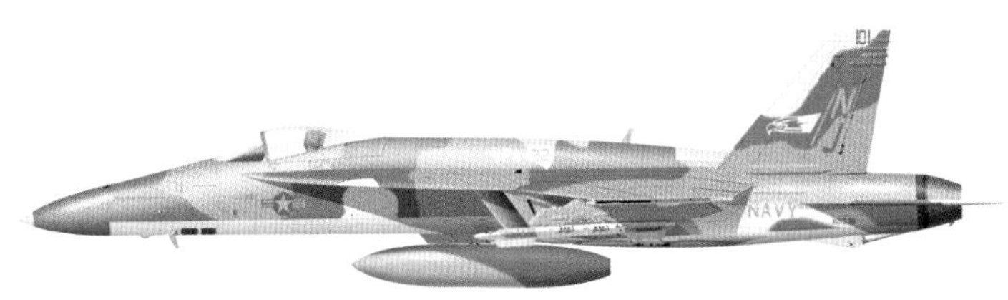
NAVY

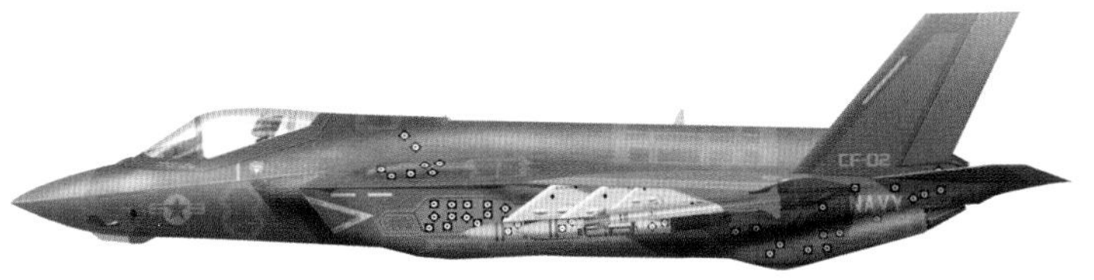
CF-02

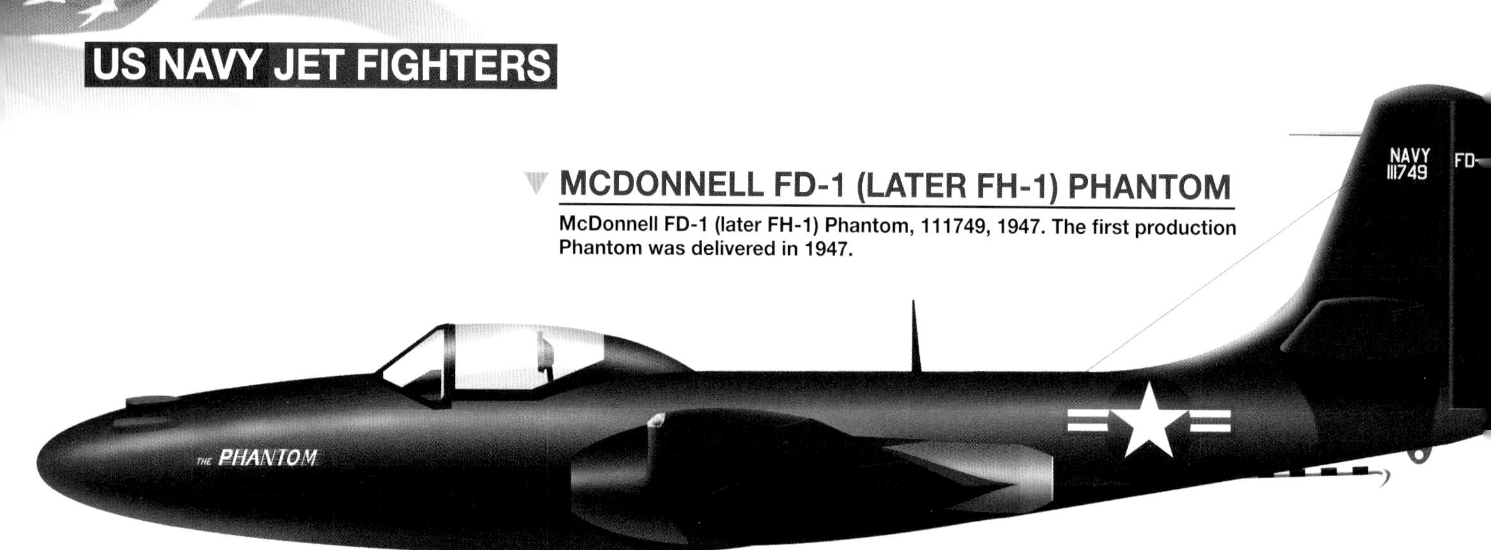

MCDONNELL FD-1 (LATER FH-1) PHANTOM

McDonnell FD-1 (later FH-1) Phantom, 111749, 1947. The first production Phantom was delivered in 1947.

MCDONNELL FH-1 PHANTOM

Primitive and produced in only small numbers, McDonnell's FH-1 Phantom was nevertheless the US Navy's first operational pure jet fighter.

1947-1954

Engineer James McDonnell's timing was impeccable. He left aircraft manufacturer Glenn L. Martin in 1938, having worked there for seven years, to form his own company making aircraft components in 1939 – just as the US military was preparing to place enormous orders for parts.

McDonnell was keen to begin building aircraft of his own and just a year later his team entered a US Army Air Corps contest to design a new high-speed long-range bomber interceptor. After a steep learning curve, the result was the XP-67 Moonbat – a novel but underpowered prototype which was destroyed following an engine fire before testing could be completed.

The fledgling firm had no experience of naval aircraft or jet propulsion but with other manufacturers working at maximum capacity and having demonstrated the ability to innovate, McDonnell received a contract to build a jet-powered fighter for the Navy in August 1943.

The result was a compact fighter with folding straight wings; tailplane with noticeable dihedral; forward-set full-vision canopy; four .50-cal machine guns in its upper nose; a tricycle undercarriage and a jet engine buried in each wingroot. It was designed with the first American turbojet in mind – the Westinghouse WE-19XB-2B, later redesignated the J30-WE-20 – which produced 1600lb of thrust.

Two prototypes were ordered and the first one made its maiden flight on January 26, 1945, although short hops on a single engine had previously been carried out. An order for 100 full production examples followed on March 7. The production model had a fuselage 18in longer than the prototypes, its rudder was shortened and its tailplanes were reduced in length. An aerodynamic external fuel tank could be installed beneath the fuselage for added range and the canopy windscreen was changed to improve pilot visibility.

The end of the Second World War saw the production order cut to just 60 examples. The second prototype became the first pure jet fighter to make an arrested landing on a US carrier on July 21, 1946, and the first production FD-1 flew on November 28. Deliveries to US Navy squadrons commenced on July 23, 1947, and the type was redesignated FH-1. Its service career would prove to be brief however, and by mid-1949 it was being replaced by McDonnell's own F2H-1 Banshee. Training and reserve units would continue to use the type until 1954.

The Phantom suffered from weak engines, a lack of firepower, an inability to carry external stores and no ejection seat but better aircraft were on the way.●

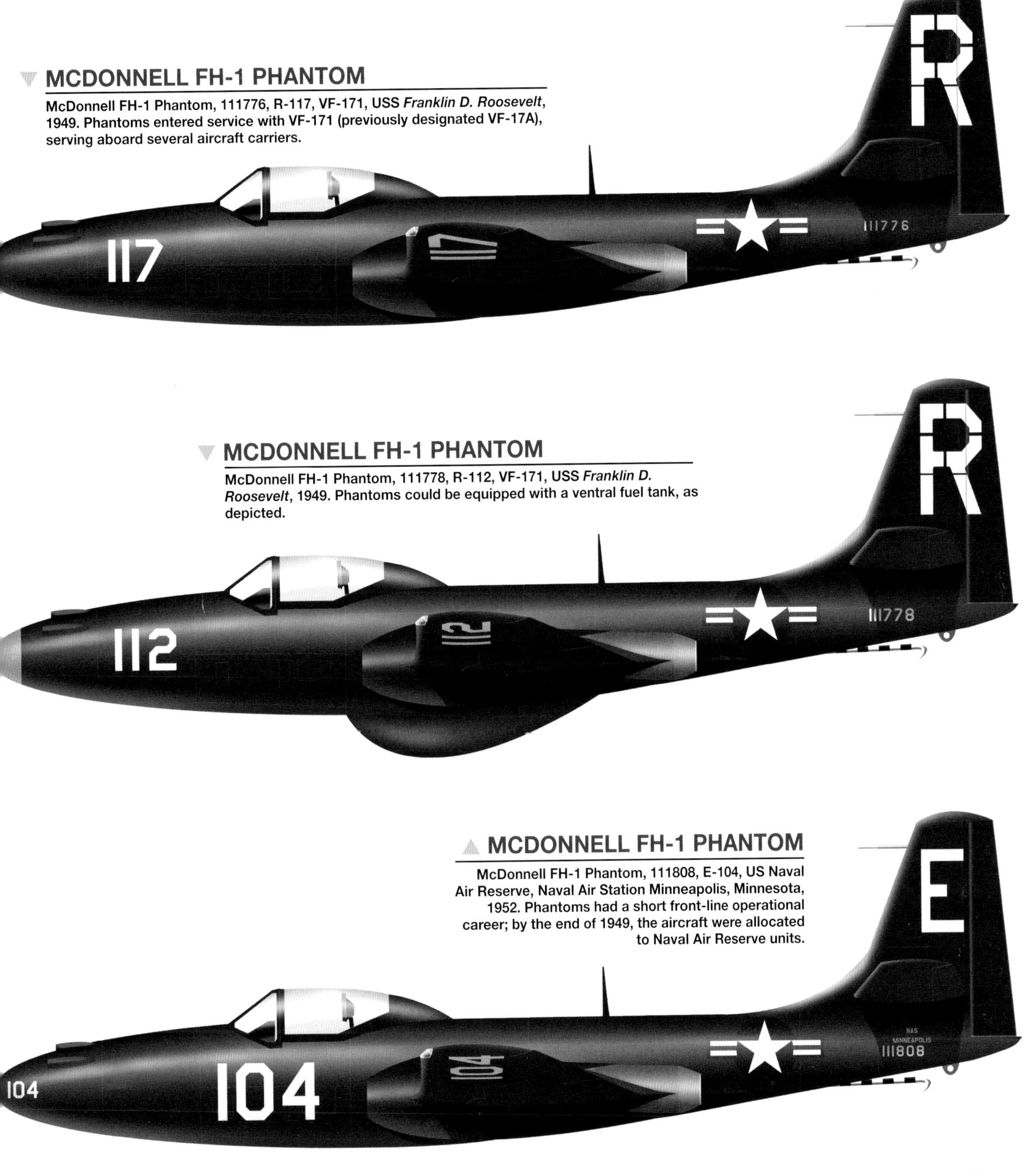

MCDONNELL FH-1 PHANTOM

McDonnell FH-1 Phantom, 111776, R-117, VF-171, USS *Franklin D. Roosevelt*, 1949. Phantoms entered service with VF-171 (previously designated VF-17A), serving aboard several aircraft carriers.

MCDONNELL FH-1 PHANTOM

McDonnell FH-1 Phantom, 111778, R-112, VF-171, USS *Franklin D. Roosevelt*, 1949. Phantoms could be equipped with a ventral fuel tank, as depicted.

MCDONNELL FH-1 PHANTOM

McDonnell FH-1 Phantom, 111808, E-104, US Naval Air Reserve, Naval Air Station Minneapolis, Minnesota, 1952. Phantoms had a short front-line operational career; by the end of 1949, the aircraft were allocated to Naval Air Reserve units.

MCDONNELL

The fundamentally sound design of McDonnell's Phantom had been let down by its engines – so as bigger and better powerplants became available it made sense to simply scale up the FH-1's airframe to accommodate them. The result was the F2H Banshee.

1948-1965

Just five days before it received the order for 100 production model FH-1 Phantoms, on March 2, 1945, McDonnell was handed a contract to build three prototypes of a scaled-up version designated the XF2D-1.

The additional size would allow the aircraft to accommodate a pair of Westinghouse J34-WE-22 turbojets in its wingroots, each producing 3000lb of thrust, and four 20mm cannon in its nose rather than four machine guns. The wings remained straight and were still capable of folding to suit the cramped confines of a hangar deck – but each now had four

MCDONNELL F2H-2 BANSHEE

McDonnell F2H-2 Banshee, F-210, VF-62, USS *Lake Champlain* (CVA-39), aboard USS *Wasp* (CV-18), 1952. The punishment for landing on the wrong carrier was swift and very visible, as seen in this aircraft.

F2H BANSHEE

external hardpoints, allowing the Banshee to carry bombs or rockets weighing up to a total of 1540lb.

McDonnell also took the opportunity to make all-over detail changes to the aircraft's aerodynamic form. The Banshee, affectionately known as the 'Banjo' in service, would effectively incorporate all the lessons learned from designing and building the Phantom.

The first prototype XF2H-1 flew on January 11, 1947, as McDonnell was gearing up to commence deliveries of the FH-1, and an order for 56 production model F2H-1s followed on May 29. And an order for the further improved F2H-2 was placed even before the first F2H-1s had been delivered. This second variant would become the most-produced Banshee with 364 examples rolling off McDonnell's St Louis production line.

The F2H-2 was powered by J34-WE-34s, producing 3250lb of thrust, had an ejection seat and could be fitted with wingtip fuel tanks for improved range. Unfortunately, the tip tanks could only be fuelled when the wings were down and had to be drained before the wings could be folded – making the Banshee more difficult to work with.

While production was ongoing it was decided that a number of Banshees should be able to carry a single Mk.7 or Mk.8 nuclear bomb beneath their port wingroot if needed. Consequently, 27 airframes received strengthened port wings, had one 20mm cannon replaced with an inflight refuelling probe and had landing gear that could be pumped up to provide additional ground clearance for the large ordnance.

These aircraft received the designation F2H-2B. Another 14 airframes were modified on the production line to become the Navy's first night and all-weather interceptors, designated F2H-2N. These aircraft were fitted with AN/APS-19A radar housed inside a new ▶

MCDONNELL F2H-2 BANSHEE

McDonnell F2H-2 Banshee, T-111, VF-11, USS *Kearsarge* (CVA-33), Korea, 1952. Banshees proved very useful during the Korean War as ground-attack aircraft.

McDonnell F2H Banshee

MCDONNELL F2H-2 BANSHEE

McDonnell F2H-2 Banshee, O-312, VA-76, NAS Oceana, Virginia, 1955. One of the last units to operate the F2H-2 was VA-76.

MCDONNELL F2H-3 BANSHEE

McDonnell F2H-3 Banshee, I-109, VF-41, USS *Forrestal* (CVA-59), 1956. The Banshees of VF-41 operated from the US Navy's first carrier to be constructed with a steam catapult, an angled flight deck and an optical landing system.

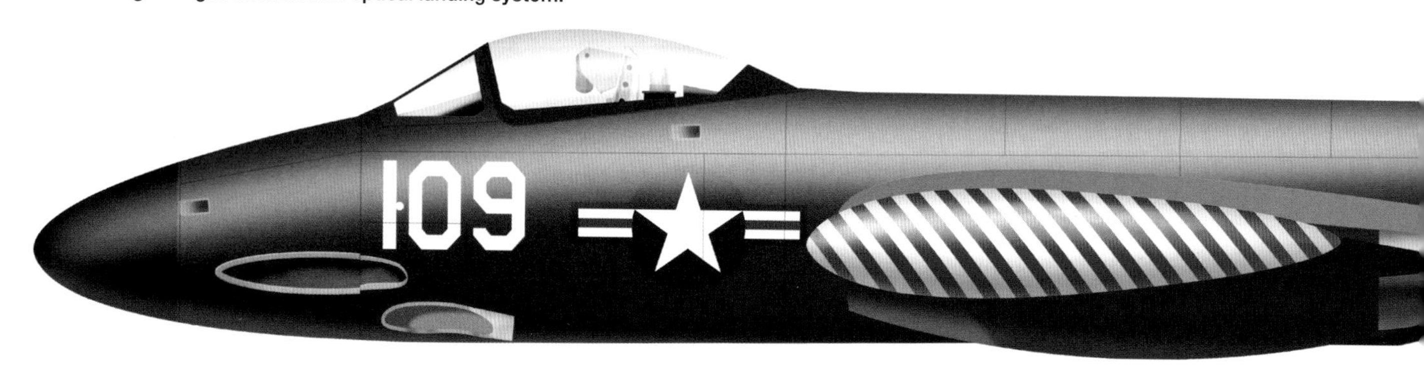

MCDONNELL F2H-3 BANSHEE

McDonnell F2H-3 Banshee, K-112, VF-31, 1955. Banshees were adapted to carry nuclear ordnance including the depicted Mk.7 bomb; this aircraft carries one such weapon under the right wing and is equipped for in-flight refuelling.

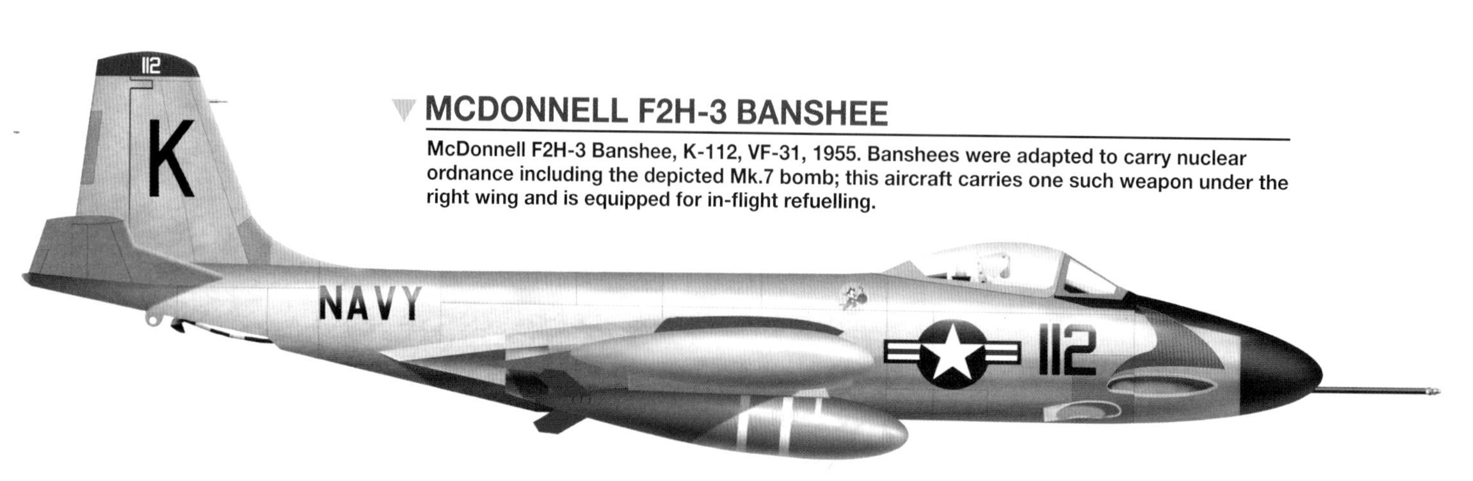

lengthened plastic nose section and had their cannon shifted further aft.

A photo reconnaissance variant was produced under a separate contract as the F2H-2P, with 58 examples being built from scratch. Again, the F2H-2 was given a much longer nose. This was divided into three sections and incorporated various windows so that cameras could be installed in a variety of different positions.

Meanwhile, McDonnell had been busy with another scaling exercise and had proposed a still larger version of the Banshee as a dedicated night and all-weather interceptor. The F2H-2N had already proven that the aircraft could be adapted to this purpose so on July 14, 1950, an order was placed for 250 F2H-3s. While the F2H-3 retained the engines of its predecessor, it had a new 7ft longer fuselage, newly enlarged wings and a completely redesigned tail unit. It maintained some parts commonality but was practically a new aircraft. The fuselage could house two new fuel tanks, which meant that tip tanks were no longer always necessary, and inside the nose was an AN/APQ-41 radar.

The last Banshee sub-type was the F2H-4 – another night and all-weather interceptor. It was based on the F2H-3 and externally looked identical. However, beneath the skin it had new J34-WE-38 engines offering 3600lb of thrust each and a new AN/APG-37 radar in its nose. One hundred and fifty examples were ordered by the Navy.

A total of 892 Banshees were built across the four variants and in September 1962 the F2H-3 was redesignated F-2C while the F2H-4 became the F-2D. The type remained in front line service until 1959 and the last examples were retired from the Reserves in 1965, ending a lineage that had begun during the dark days of the Second World War.●

MCDONNELL F2H-4 BANSHEE

McDonnell F2H-4 Banshee, X-102, VF-102, USS *Randolph* (CVA-15), 1956. The F2H-4 variant was procured in relatively low numbers but did feature a new radar giving true all-weather capability

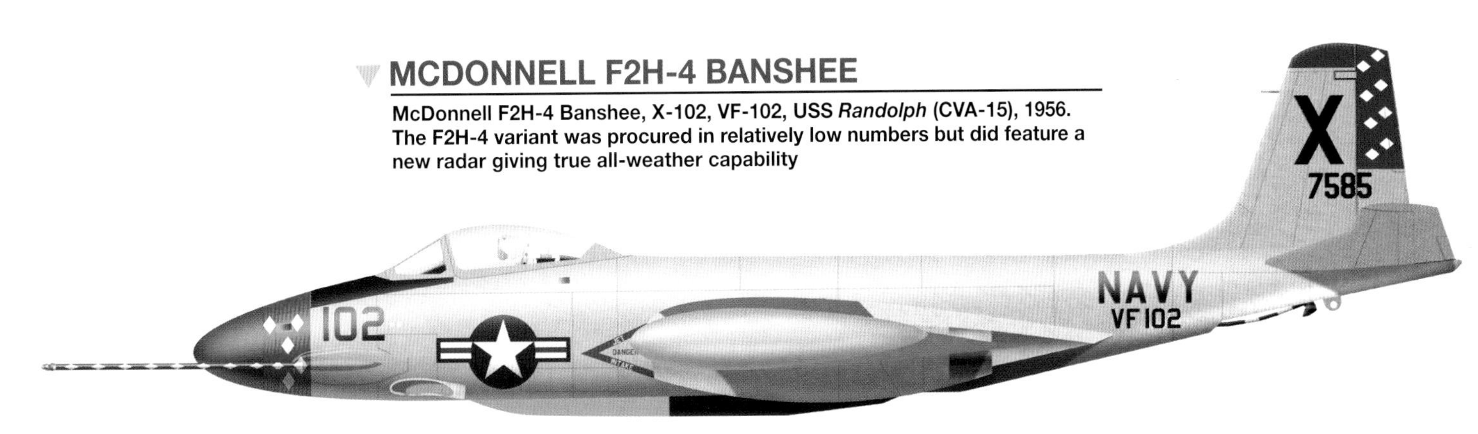

NORTH AMERICAN FJ-1 FURY

The FJ-1 Fury was a 'safe' single-jet design produced at a time of uncertainty about the advantages and potential perils of swept-wing design. It was also a snapshot of the F-86 during the early stages of its development.

1947-1953

NORTH AMERICAN FJ-1 FURY

North American FJ-1 Fury, F-103, NAS Oakland, California, 1952. The operational career of the FJ-1 was short, with aircraft being assigned to reserve units and eventually retired in 1953.

As McDonnell's Phantom approached the prototype testing phase it became clear that the Second World War was nearing its conclusion. The team behind the famous Mustang had therefore been at liberty to work up a jet fighter design and the US Navy ordered three examples of it.

North American Model NA-134 was, like the Phantom, a straight-winged machine – but unlike the Phantom it was powered by only a single large turbojet, the General Electric J35-GE-2. It had three .50-cal machine guns on either side of the nose intake, a tricycle undercarriage and the ability to fit wingtip tanks. It lacked folding wings but had a system allowing the nosewheel leg to retract into the nose, so that the aircraft could 'kneel'. Its nose could then be tucked under the tail of the aircraft in front – saving space on the hangar deck.

The Navy gave North American's design the official designation XFJ-1 and during its development the company offered to incorporate the latest advances in aerodynamics, including swept wings. The Navy declined but had placed an order for 100 Furies. The first prototype took its maiden flight on September 11, 1946, and deliveries of the production model commenced on November 27, 1947.

However, it quickly became evident that the FJ-1 was not well suited to carrier operations and production was cut short after just 30 had been built. VF-51 commenced operations with eight brand new Furies during August 1948 aboard the USS *Princeton* – but within 48 hours all eight had been wrecked in crashes or were so badly damaged that they could no longer fly.

The FJ-1 managed just 14 months with VF-51, the only active Navy fleet squadron to operate the type, before those aircraft that remained were transferred to the Reserve.

While the first Fury was unsuccessful, North American was undeterred and developed the design into the USAF's F-86 Sabre which proved highly successful during the Korean War. The 'Fury' name would return when navalised versions of the Sabre were devised.●

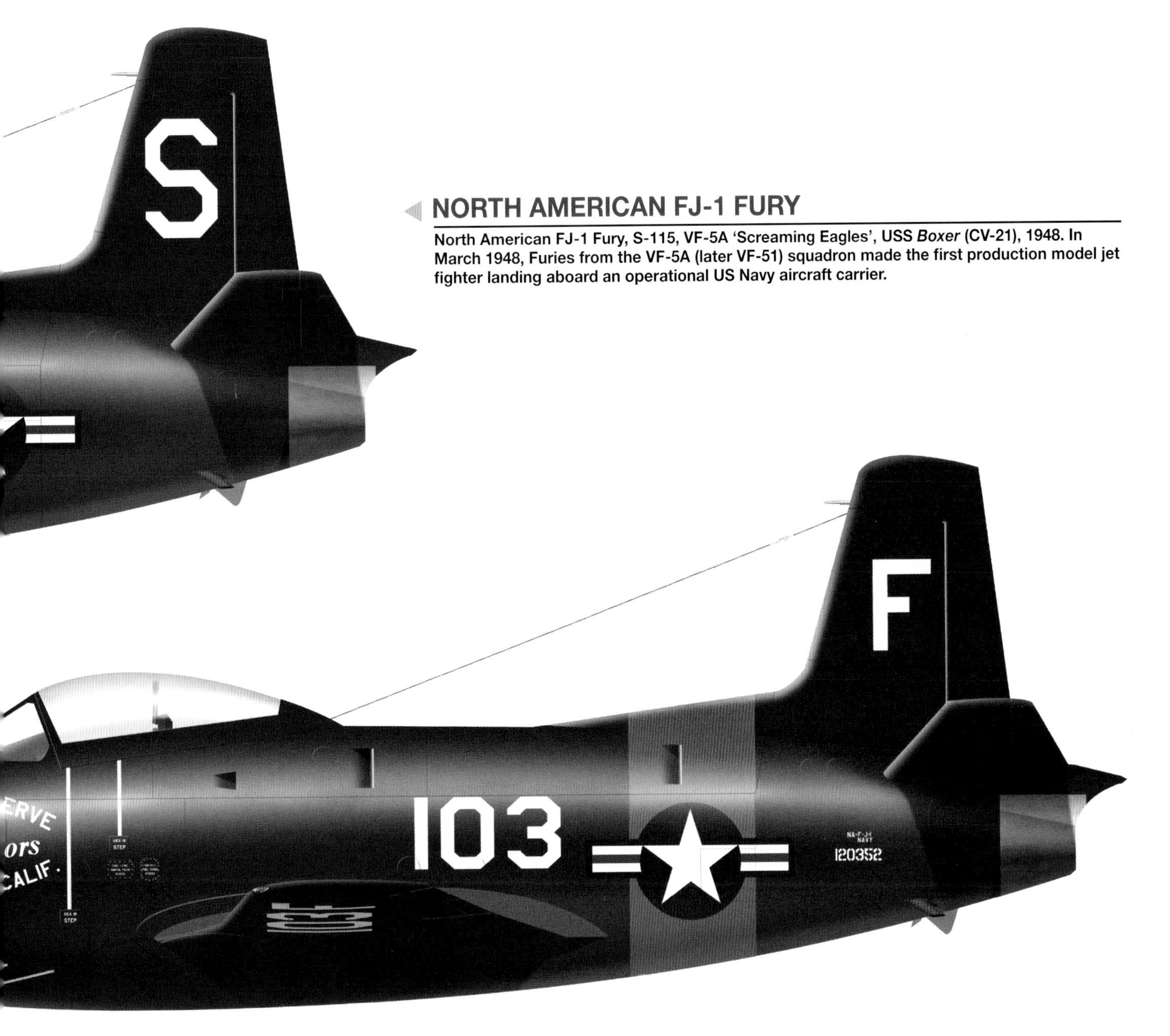

NORTH AMERICAN FJ-1 FURY

North American FJ-1 Fury, S-115, VF-5A 'Screaming Eagles', USS *Boxer* (CV-21), 1948. In March 1948, Furies from the VF-5A (later VF-51) squadron made the first production model jet fighter landing aboard an operational US Navy aircraft carrier.

NORTH AMERICAN FJ-2 AND FJ-3 FURY

The FJ-1 had been a failure as a carrier aircraft but it had provided North American with useful experience while developing the USAF's F-86. This in turn benefitted the USN when it bought a navalised version of the Sabre.

NORTH AMERICAN FJ-2 FURY

North American FJ-2 Fury, FT-930, NATC, USS *Coral Sea*, 1953. The FJ-2 Fury was based on the F-86E, featuring several modifications for carrier operations.

1954-1962

Tendering for a USAAF fighter contract in 1945, North American took the design it had already offered to the US Navy – which became the FJ-1 Fury – and deleted the Navy's very specific requirements from it. Unburdened by the Navy's strength, landing and take-off needs, the design could be much lighter and therefore faster.

The result was the superb F-86 Sabre, which entered service with the USAF in 1949, two years after the Fury's brief career with the Navy had begun. The success of the F-86 prompted the Navy to request a navalised variant. North American took Model NA-170, the F-86E, and added a longer nosewheel leg, an arrestor hook under the rear fuselage, a catapult bridle hook and folding wings. An order for three prototypes of this design was placed on March 8, 1951, under the designation XFJ-2.

Converting standard F-86Es was straightforward, since the prototypes were not required to have folding wings, and the first flew on December 27, 1951. Following trials, the aircraft underwent undercarriage and arrestor gear strengthening.

Three hundred production FJ-2s had been ordered initially but this was cut to 200 when the Korean War ended. When the first production FJ-2s were delivered towards the end of 1952, however, it was discovered that adding all the carrier equipment and strengthening had removed much of the performance advantage enjoyed by the F-86 and actually made the aircraft unsuitable for

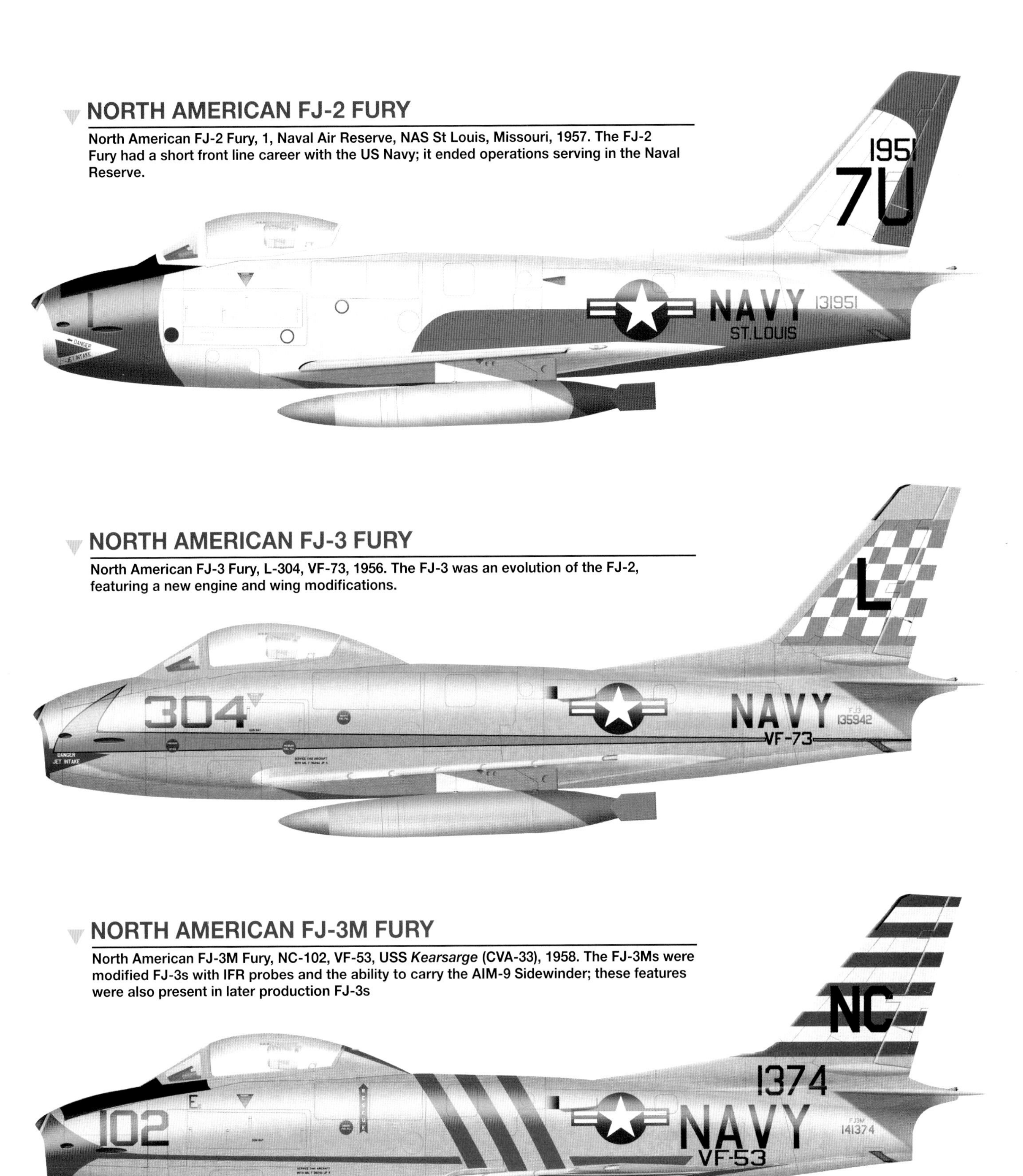

NORTH AMERICAN FJ-2 FURY

North American FJ-2 Fury, 1, Naval Air Reserve, NAS St Louis, Missouri, 1957. The FJ-2 Fury had a short front line career with the US Navy; it ended operations serving in the Naval Reserve.

NORTH AMERICAN FJ-3 FURY

North American FJ-3 Fury, L-304, VF-73, 1956. The FJ-3 was an evolution of the FJ-2, featuring a new engine and wing modifications.

NORTH AMERICAN FJ-3M FURY

North American FJ-3M Fury, NC-102, VF-53, USS *Kearsarge* (CVA-33), 1958. The FJ-3Ms were modified FJ-3s with IFR probes and the ability to carry the AIM-9 Sidewinder; these features were also present in later production FJ-3s

carrier operations. With better aircraft, such as the F9F-6 Cougar, becoming available the FJ-2 only lasted in front line service until 1956.

North American had the answer, however. Where the FJ-2 had the 6000lb thrust General Electric J47-GE-2 engine, the FJ-3 was designed to accommodate the 7220lb thrust Wright J65-W-2 engine. The fifth production FJ-3 was modified to take the new engine and first flew with it on July 3, 1953. A total of 389 FJ-3s were ordered and deliveries commenced in March of the following year. A second order for 214 followed, later reduced to 149, for a total of 538 FJ-3s. The second batch came with the even more powerful J65-W-4D, which had 7660lb of thrust.

Both FJ-2 and FJ-3 were armed with four 20mm cannon but the latter could take an external load of drop tanks, bombs or rockets. The last 80 FJ-3s off the line were equipped to carry the AIM-9A Sidewinder missile and became FJ-3Ms.

The last FJ-3s were phased out of Reserve units in 1962.●

NORTH AMERICAN FJ-4 FURY

The fourth and final Fury was the last model in the Fury/F-86 lineage and a substantial redesign pushed the aircraft's original form about as far as it could go.

NORTH AMERICAN FJ-4 FURY

North American FJ-4 Fury, Columbus, Ohio, 1954 – one of the first FJ-4 Fury prototypes, unpainted.

North American's replacement for the F-86, the F-100 Super Sabre, made its first flight on May 25, 1953, but the Navy wasn't yet finished with its predecessor. The FJ-3 Fury had proven successful and North American was offering a thoroughly upgraded and improved version that built on all the experience accumulated up to that point – a safer bet than taking a whole new type.

As a result, on October 16, 1953, the Navy placed an order for two prototypes of the FJ-4. Although nominally another Fury, it was practically a new aircraft with a new fuselage, new wings and a new tail. The fuselage had new contours and the cockpit was set lower down, the bigger wide-chord wings folded further outboard and allowed increased internal fuel storage.

The landing gear was repositioned, the tailplanes were thinner and an uprated J65-W-16A engine produced a healthy 7700lb of thrust. Up to 2000lb of stores could be carried externally on four underwing pylons, including 150 gallon tanks, Sidewinder rails, rocket pods and bombs. A refuelling probe could also be fitted beneath the port wing.

The first prototype flight was on October 28, 1954, and a total of 150 production model FJ-4s were ordered. Deliveries commenced four months later but nearly every aircraft went to a Marine squadron.

Rocketdyne, a division of North American, converted two FJ-4s into FJ-4Fs by installing AR-1 rocket motors in their tails – a programme intended to aid development of the F8U-3 Crusader III – before the final version of the FJ-4 was brought into service, the FJ-4B. The latter was a fighter-bomber variant with strengthened wings capable of carrying tactical nuclear weapons. It had six underwing hardpoints rather than four and could carry a wide range of stores – including a buddy refuelling system which turned the FJ-4B into a aerial tanker. Extra speed brakes were also added to the aft fuselage and the landing gear was lengthened.

The first production FJ-4B flew on December 3, 1956, and deliveries of another 221 were made up to May 1958. The last FJ-4Bs were phased out in the mid-1960s.●

NORTH AMERICAN FJ-4B FURY

North American FJ-4B Fury, XF-14, VX-4, NAS Point Mugu, California, 1960. This example carries the ASM-N-7 Bullpup air-to-surface missile.

NORTH AMERICAN FJ-4B FURY

North American FJ-4B Fury, NP-210, VA-216, 1960. The FJ-4B could carry a tactical nuclear bomb, like this MK.7.

NORTH AMERICAN FJ-4 FURY

North American FJ-4 Fury, HU-35, VU-7, Naval Air Station Brown Field, California, 1960. FJ-4 Furies were operated by Naval Air Reserve units until the mid-1960s.

DOUGLAS F3D

1951-1970

The slow and underpowered Skyknight night-fighter's side by side seating arrangement and generous accommodation for electronic equipment gave it a longevity enjoyed by few other early 1950s fighters.

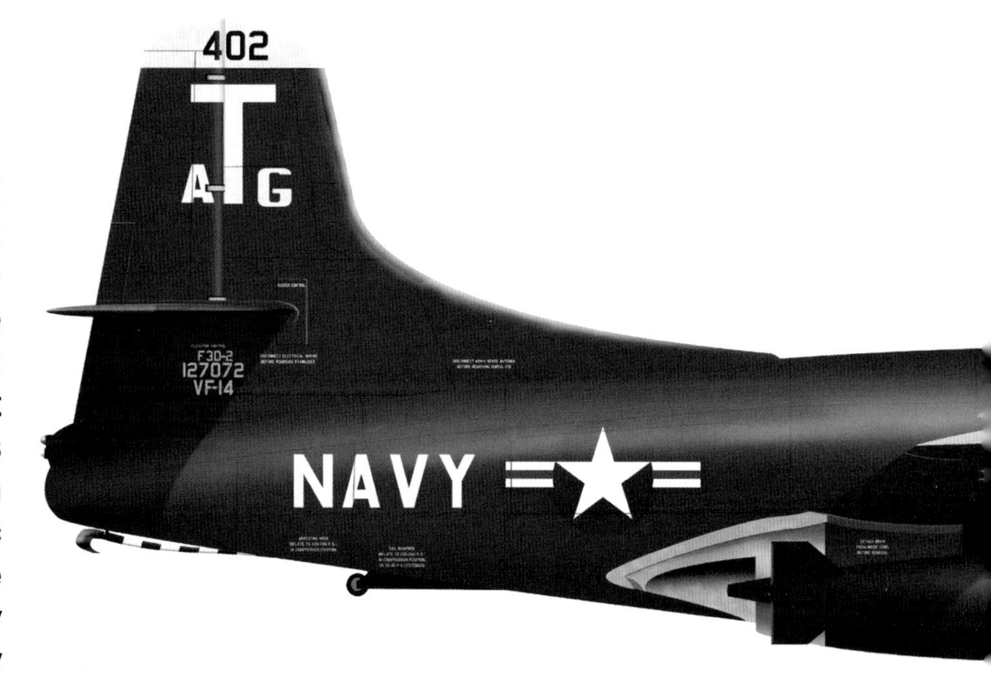

DOUGLAS F3D-2 SKYKNIGHT

Douglas F3D-2 Skyknight, NA-603, VC-4, USS *Franklin D. Roosevelt* (CVB-42), Korea, 1953. VC-4 was a composite squadron operating both piston and jet aircraft, providing detachments to several aircraft carriers and to land bases during the Korean War.

SKYKNIGHT

The design that would eventually become Douglas's F3D Skyknight was begun in 1945 after the Navy issued a requirement for a jet night fighter with two crew positioned side by side in a comfortable pressure cabin. Its chief competitor was a Grumman design known as the XF9F-1.

The Skyknight's broad fuselage allowed for a massive nose-mounded radome as well as four cannon positioned beneath it. Its unswept wings were mid-mounted and each had two hardpoints capable of carrying bombs and rockets – even though the aircraft was designed as a night fighter. Its engines were housed in bulky nacelles attached directly to the fuselage and a smaller radome was positioned at the tip of its tail. This enclosed a radar system that would warn the crew if an enemy aircraft was approaching from behind.

The rear fuselage incorporated large air brakes which were similar to those installed on Douglas's successful Skyraider, just being introduced in 1946. While the cockpit was comfy, getting into and out of it was difficult. Getting in involved scaling the aircraft's right hand side and dropping down through an opening in the upper canopy. Getting out involved following this same procedure in reverse.

DOUGLAS F3D-2 SKYKNIGHT

Douglas F3D-2 Skyknight, T-402, VF-14, USS *Intrepid* (CVA-11), 1954. VF-14 operated the Skyknight for only about a year.

Douglas F3D Skyknight

DOUGLAS F3D-2 SKYKNIGHT

Douglas F3D-2 Skyknight, NAS Point Mugu, California, 1956. Despite a rather short frontline career by the US Navy, Skyknights were used for several years by test and evaluation units; 124610 was fitted with the radome and radar of the F-4 Phantom.

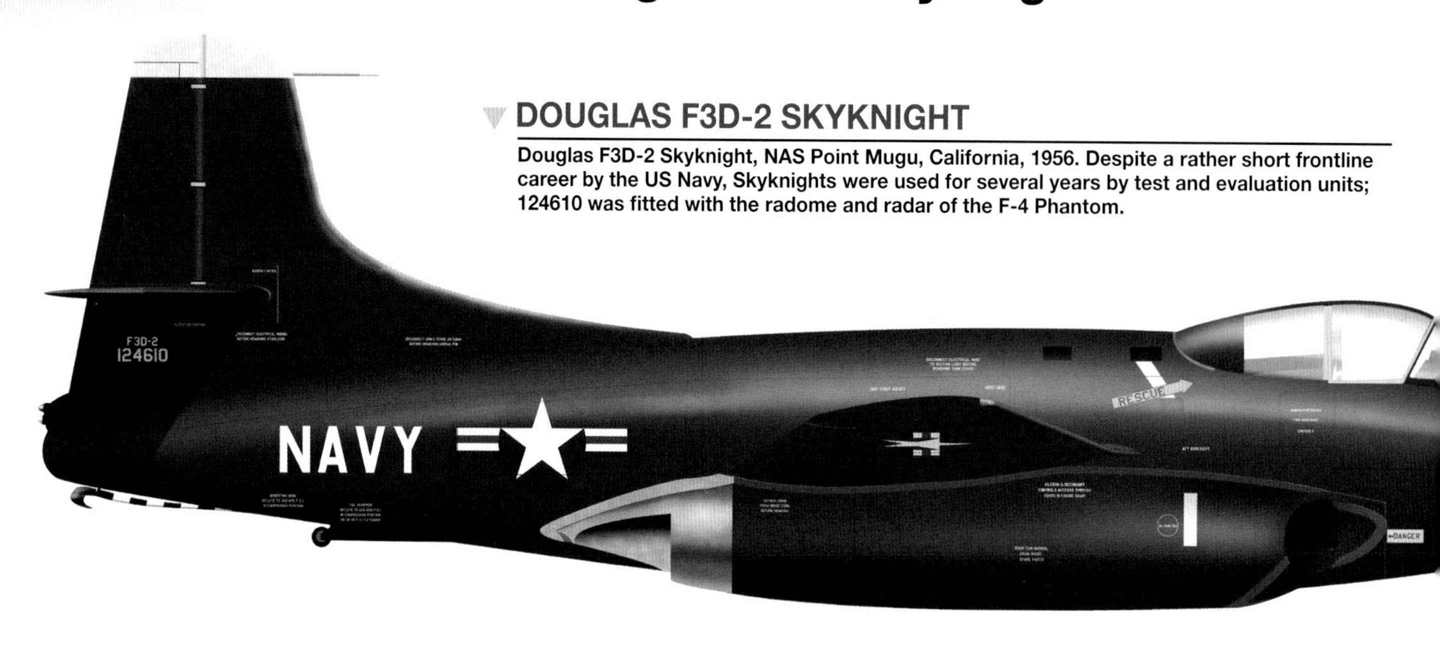

DOUGLAS F3D-2 SKYKNIGHT

Douglas F3D-2 Skyknight, AD-178, VF-101, NAS Key West, Florida, 1958. The Skyknight was operated by the fleet replacement squadron VF-101 to train crews for the type.

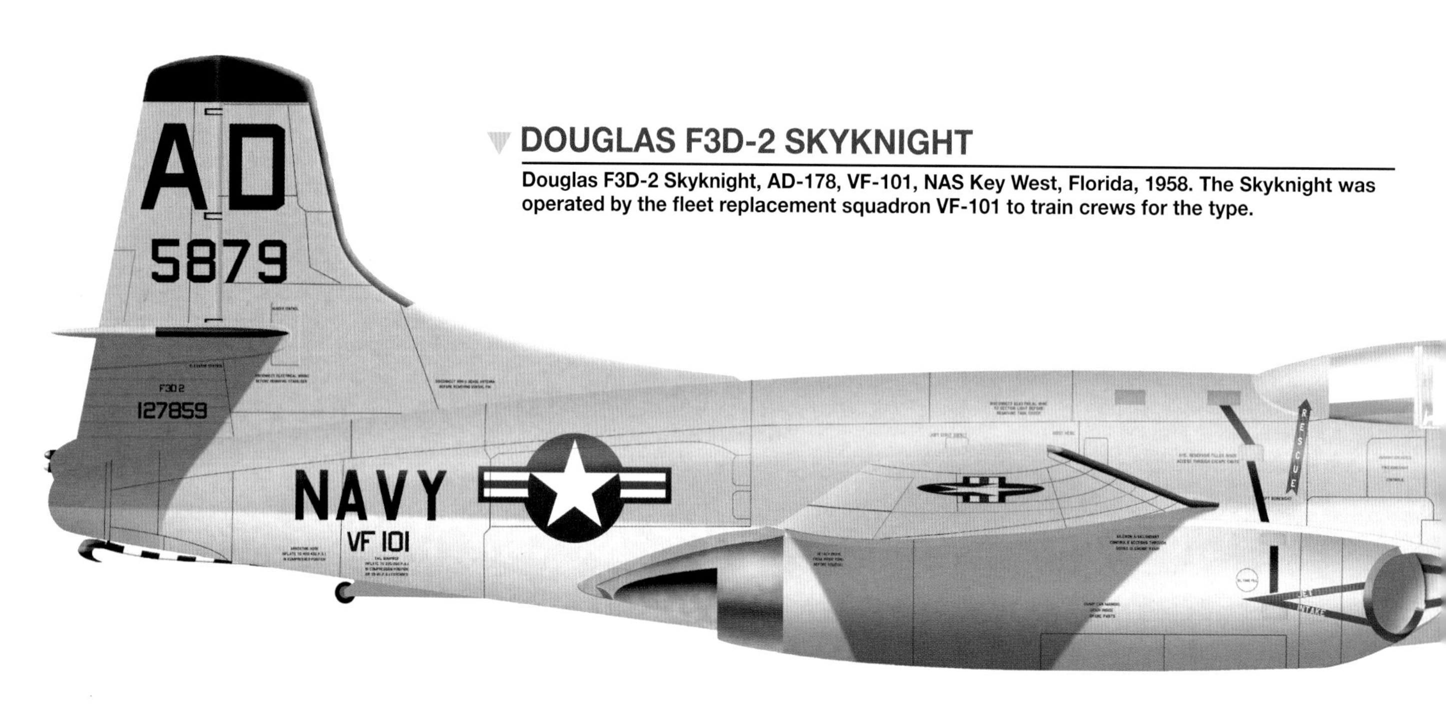

DOUGLAS F3D-2 SKYKNIGHT

Douglas F3D-2 Skyknight, 758, NAS Point Mugu, California, 1958. Skyknights were used in several air-to-air missile test programmes, including for the AAM-N-3 Sparrow II, depicted here.

In the event of an in-flight emergency, there were no ejection seats. Instead, escaping the Skyknight involved clambering down into a tunnel which led to a hatch on the lower fuselage, opening it, then crawling out into open air before manually opening a parachute.

The Navy ordered three XF3D-1 prototypes on April 3, 1946, and work progressed in parallel with Grumman's work on the XF9F-1 – a design with a long slender fuselage and two engines in a single nacelle on each wing.

The Grumman design was cancelled six months later and the first XF3D-1 flew on March 23, 1948, powered by 3000lb thrust Westinghouse J34-WE-22s. A production order for just 28 F3D-1s eventually followed on June 26, 1949, and the first of these flew on February 13, 1950 with J34-WE-32s, which were more powerful at 3250lb thrust each but required bigger nacelles.

A dozen F3D-1s were redesignated F3D-1Ms after being fitted out to carry and launch the Sparrow III missile – but this was found to be very unreliable and almost impossible to use against targets which did anything other than fly in a straight line.

The initial short run of Skyknights was followed by a much longer series – the F3D-2. This was powered by two J34-WE-36/36As, producing 3400lb of thrust each. Weaponry remained the same but up to 4000lb of stores could be carried externally.

While the F3D-2 was qualified for carrier operations, it never made a full carrier deployment. The only two Navy fleet squadrons to operate it were VF-11 and VF-14. Skyknights were operated in Korea by VMF(N)-513, Marine unit, beginning in June 1952. Though only a handful were ever operated at one time, they performed well as night fighters and successfully destroyed around half a dozen enemy aircraft. They were also successfully used to escort B-29s during night missions.

Sixteen F3D-2s were modified for Sparrow I testing as F3D-2Ms and a single F3D-2 was used for other special equipment testing as an F3D-2B. A number of F3D-2s became night-fighter trainers under the designation F3D-2T and others F3D-2T2s as radar operator trainers.

Although its performance had been poor from the beginning – a top speed of 426mph at 15,000ft for even the F3D-2 – the Skyknight's side by side seating arrangement ensured its survival. A total of 35 were converted into F3D-2Qs for electronic warfare operations. The equipment they carried enabled them to locate enemy transmitters, jam them and deploy countermeasures.

The type was redesignated F-10 in September 1962, with F3D-1s becoming F-10As, F3D-2s becoming F-10Bs, F3D-2Ms becoming MF-10Bs, the trainers becoming TF-10Bs and the F3D-2Qs becoming EF-10Bs.

VMCJ-1 flew both EF-10Bs and RF-8A Crusaders from Da Nang Air Base during the Vietnam War, starting in April 1965. At least one EF-10B was shot down by a North Vietnamese SA-2 surface-to-air missile.

Even after the Skyknight was retired from active service in 1970, a trio of airframes served with the US Army gathering telemetry data into the early 1980s.●

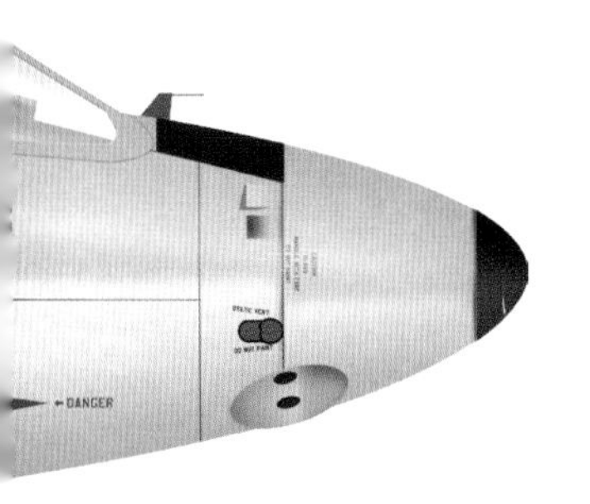

DOUGLAS F3D-2T-2 SKYKNIGHT

Douglas F3D-2T-2 Skyknight, NJ-197, VF-121, NAS Miramar, California, 1961. The F3D-2T2 variant was used to train radar-operators and as an electronic warfare aircraft.

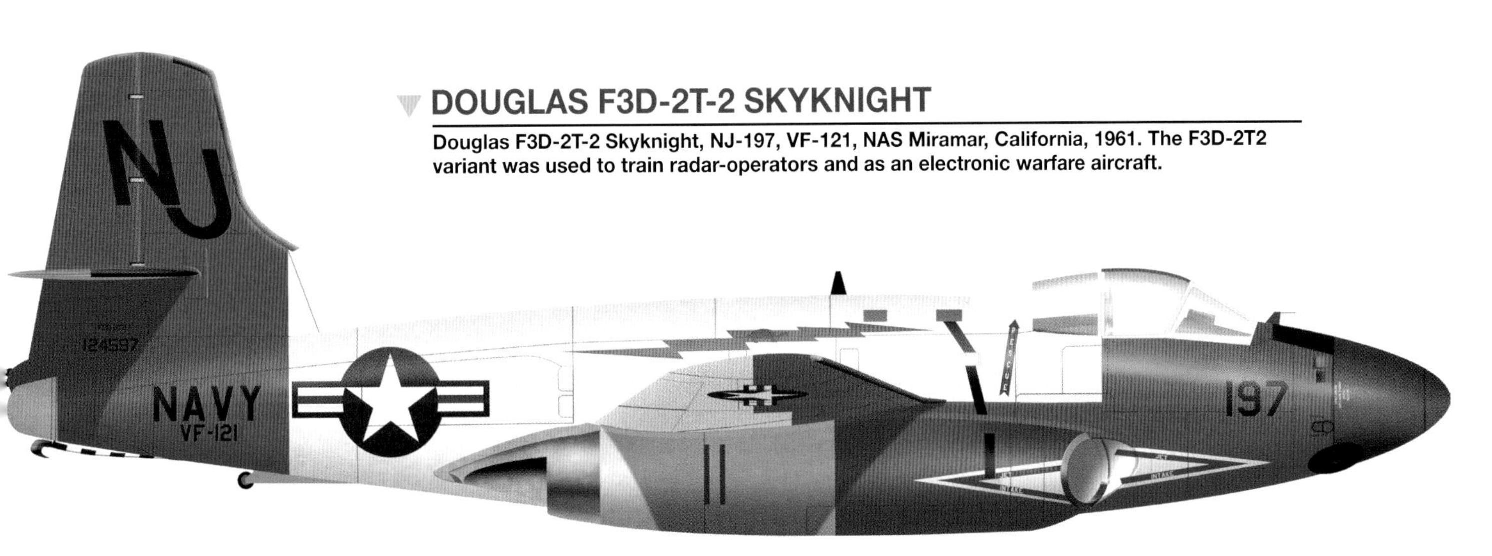

GRUMMAN F9F

GRUMMAN F9F-3 PANTHER

Grumman F9F-3 Panther, S-109, VF-51, USS *Valley Forge* (CV-45), 1950. The first US Navy kill by a carrier-based jet fighter was achieved by Lt (jg) Leonard Plog, flying this aircraft when it shot down a North Korean Yak-9. VF-51's squadron emblem (below the cockpit) is sometimes misidentified as a kill mark; later a kill mark was added to the aircraft for publicity photos.

1949-1958

Grumman's F9F stood head and shoulders above the Navy's other late 1940s/early 1950s jet fighter designs. It was reasonably powerful, moderately adaptable and capable of packing a decent punch in combat.

GRUMMAN F9F-2B PANTHER

Grumman F9F-2B Panther, A-122, VF-721, USS *Boxer* (CV-21), 1951. F9F-2Bs could be equipped with rockets and bombs for air-to-ground missions; this aircraft displays impressive mission markings.

PANTHER

Strangely enough the first F9F, the XF9F-1, was a four-engined two-seater night-fighter designed by Grumman to meet the same specification as the Douglas F3D Skyknight. The aircraft was never built – being defeated by the Skyknight on October 9, 1946.

Grumman took the F9F back to the drawing board and split the project, known as Design 79, into four different strands. Two had composite propulsion – with both turbojets and rockets – one had a turbojet in each wing and the last was powered by a single turbojet.

This fourth design, known as 'Study D', appealed to the Navy and was chosen for further development. Three prototypes were ordered, two as XF9F-2s powered by Pratt & Whitney's 5000lb thrust J42-P-6 (a licence-built version of the British Rolls-Royce Nene, which could achieve 5700lb thrust with water injection) and one as an XF9F-3, powered by the American designed and built Allison J33-A-8 with 4600lb thrust.

The first XF9F-2 flew on November 24, 1947, while the first XF9F-3 flew nearly 10 months later on August 16, 1948. The Navy decided to hedge its bets by ordering 47 F9F-2s and 54 F9F-3s. The first F9F-3s were delivered in May 1949 but it was soon found that the F9F-2 offered superior performance and the Allison design was plagued with technical issues. Most F9F-3s therefore had their Allison engines replaced with J42-P-8s and the remainder of the F9F-3 contract was reverted to F9F-2 production. The J42-P-8 was fitted to most F9F-2s – it had the same thrust as the J42-P-6 but a different ignition system.

Grumman called the new type 'Panther' – which was in keeping with the company's policy of giving its fighters feline names. It was of all-metal construction and had mid-mounted straight wings with flaps on both the leading and trailing edges. Straight tailplanes were positioned high up on the tail fin and the centrally-mounted turbojet exhausted beneath it. The main engine intakes were in the aircraft's wingroots but there was also a pair of spring-loaded auxiliary intake doors on the aircraft's back. These were used only for take-off and in the event that emergency airflow was required.

The landing gear was a tricycle arrangement; the wing-mounted main wheels retracted towards the fuselage while the nosewheel retracted rearwards. Just behind the nosewheel door was a split perforated airbrake and there was a 'stinger' type arresting hook underneath the jet exhaust.

The wings folded hydraulically just outboard of the main undercarriage but unlike many naval types they could not be folded vertically or beyond and instead still stuck out slightly on either side. Early Panthers lacked wingtip tanks but from the 13th production example on, 120 gallon tip tanks became a permanent ▶

Grumman F9F Panther

part of the aircraft's design, taking total fuel capacity to 923 gallons.

An ejection seat was provided for the pilot who sat within a pressurised cabin beneath a full-vision canopy. This could easily be slid to the rear when entering or leaving the aircraft and there was even a small retractable step provided in the lower fuselage to help the pilot get in and out.

The nose of the aircraft could be slid forwards to allow access to the cannon while the tail could be removed as a single piece to allow easy access to the engine.

Armament was four 20mm M3 cannon with 190 rounds each, fitted within the lower part of the nose, and a Mark 8 computing optical gunsight was provided for the pilot. There were initially three hardpoints beneath each wing which could be used for bombs or rocket launchers. A fourth hardpoint was later added to each wing and a Mk.51 rack could be added which allowed the aircraft to carry a 1000lb bomb.

Panthers modified to accommodate the rack became F9F-2Bs to start with but eventually all F9F-2s were modified and the 'B' was dropped. F9F-2s could manage a load of up to 2800lb, though the aircraft struggled to make a carrier take-off fully loaded. The Navy's display team, the Blue Angels, swapped their F8F Bearcats for F9F-2s during mid-1949, staging their first Panther show minus wingtip tanks on August 20.

A single F9F-3 was fitted with a new nose housing an Emerson turret. At first glance it looked very similar to a regular Panther nose but closer inspection revealed long slots housing two pairs of .50-cal machine guns which could be rotated around inside the fuselage. These weapons proved too hard to aim, however, and the experimental installation was soon removed.

The F9F-2 became the first US Navy jet fighter to see combat when VF-51, embarked upon USS *Valley Forge*, commenced operations over North Korea on July 30, 1950 – shooting down two Yak-9s.

During the war as a whole, Panthers would shoot down a total of seven MiG-15 fighters but they were primarily used as fighter-bombers against ground targets. There was also a need for a photo reconnaissance variant and a number of F9F-2s were modified as F9F-2Ps on the production line with a new nose which included windows and mounts for vertical and side-looking cameras.

F9F-4 AND F9F-5

Two further variants of the Panther had been ordered in 1949 which mirrored the engine options which had resulted in the F9F-2 and F9F-3. The F9F-4 was to have an Allison J33-A-16 with 6950lb of thrust when water injection was used, while the F9F-5 would be fitted with the Pratt & Whitney J48-P-6/-8 ▶

GRUMMAN F9F-2 PANTHER

Grumman F9F-2 Panther, M-415, VF-24, USS *Boxer* (CV-21), 1952. VF-24 was initially deployed to the Korean War in 1951 equipped with F4U Corsairs; it then transitioned to Panthers and returned to operations in Korea in early 1952.

GRUMMAN F9F-2 PANTHER

Grumman F9F-2 Panther, F-62, ATU-206, Naval Air Station Pensacola, Florida, 1956. Besides front-line service, Panthers were used to provide advanced pilot training.

GRUMMAN F9F-2 PANTHER

Grumman F9F-2 Panther, 5, Blue Angels, 1953. The Panther was the first jet used by the famous Blue Angels, from 1949 to 1954, using both the -2 and -5 variants.

GRUMMAN F9F-2KD PANTHER

Grumman F9F-2KD Panther, ZZ-34, GMGRU-1, NAS Point Mugu, California, 1958. A number of photo-reconnaissance F9F-2Ps were modified as drone-controller aircraft and received the designation F9F-2KD.

5
ZZ
F9F-2KD
123071
NAVY
GMGRU-I

Grumman F9F Panther

(a licence-built version of the British Rolls-Royce Tay engine) producing 7250lb of thrust with water/alcohol injection.

This situation played out almost exactly as it had done before – the Allison design suffered from numerous difficulties while development of the Pratt & Whitney-engined Panther proceeded relatively smoothly. As a result, the first F9F-5 made its maiden flight on December 21, 1949, while the F9F-4 had to wait until July 6, 1950. Deliveries of the F9F-5 commenced just four months later.

In all just 109 F9F-4s were built compared to 616 F9F-5s.

While they had different engines, the F9F-4 and F9F-5 shared an upgraded airframe with a fuselage which was 19in longer – the extra length being added forward of the engine intakes to provide additional internal fuel capacity. The upgraded Panther also had a taller tail and redesigned intakes.

The total weight of external stores was increased to 3465lb. Without this extra weight however, the F9F-5 could now fly faster than 600mph at sea level and its rate of climb was almost 6000ft per minute.

The F9F-5 became the definitive Panther and 36 airframes became F9F-5P photo recon variants. These differed from the earlier F9F-2P in having noses that were 12in longer so that bulkier camera equipment could be fitted. These aircraft were further altered so that no external stores could be carried except for 150 gallon external tanks on the inboard underwing pylon. They were also fitted with a General Electric G-3 autopilot unit.

Not long after the Korean War came to an end, with more capable aircraft becoming available, most Panthers were quickly relegated to service with the Reserve. However, there were still plenty of serviceable Panthers around which could be converted into drones and drone controllers for target practice. The former were given the new designation F9F-5K and the latter F9F-KD. Many Panthers remained when the new aircraft designation system was introduced in 1962, with the F9F-4 becoming the F-9C, the F9F-5 the F-9D, the F9F-5P the RF-9D, the F9F-K the QF-9D and finally the F9F-5KD becoming the DF-9E.

Only one other country used the Panther – Argentina. It bought 24 refurbished F9F-2s in 1958 and the Argentine Navy operated them from ground bases. Four of these were destroyed on the ground during a coup in 1963 and the remainder served until 1969 when they were grounded due to a lack of spare parts.

The Panther had demonstrated its worth during the Korean War and proven that it was among the best of America's first generation jet fighters. Newer designs with more advanced features were being developed that would soon supplant it – but there remained life in the old cat yet as the F9F-6 Cougar.●

GRUMMAN F9F-5 PANTHER

Grumman F9F-5 Panther, M-202, VF-63, USS *Yorktown* (CVA-10), 1953. High-velocity aerial rockets were a very effective weapon for air-to-ground Panther missions.

GRUMMAN F9F-5 PANTHER

Grumman F9F-5 Panther, H-312, VF-153, USS *Princeton* (CV-37), 1953.
The rather strange appearance of this aircraft came, allegedly, as the result of joining two halves of different aircraft into a 'new' one; this aircraft is said to have performed operational missions in the Korean War before being returned to the United States for rebuilding.

Grumman F9F Panther

GRUMMAN F9F-5 PANTHER

Grumman F9F-5 Panther, V-112, VF-111, USS *Boxer* (CVA-21), 1953. VF-111 deployed to the Korean theatre of operations in 1953, assigned to Air Task Group 1 (ATG-1).

GRUMMAN F9F-5 PANTHER

Grumman F9F-5 Panther, D-106, VF-781 USS *Oriskany* (CV-34), 1952. This aircraft was flown by Lt Royce Williams – who was credited with four MiG-15 kills during a single mission on November 18, 1952; due to battle damage the aircraft was ditched off the carrier at the end of the mission.

GRUMMAN F9F-5 PANTHER

Grumman F9F-5 Panther, 043, NOTS China Lake, California, 1955. The Naval Ordnance Test Station used several Panthers to perform weapons experiments; this aircraft carries four underwing 5in rockets pods.

GRUMMAN F9F-5KD (DF-9E) PANTHER

Grumman F9F-5KD (DF-9E) Panther, UA-35, VU-1, NAS Barbers Point, Hawaii, 1964. Some F9F-5Ps were modified to serve as drone controllers with the designation F9F-5KD. Several of these were later used as drones themselves, designated DF-9E.

GRUMMAN F9F-6 COUGAR

Grumman F9F-6 Cougar, M-112, VF-24, USS *Yorktown* (CV-10), 1953. VF-24 was the first squadron to deploy the new F9F Cougar; it was sent to Korea but did not arrive in time to see combat.

GRUMMAN F9F

1952-1974

Swapping the Panther's straight wings and tailplanes for swept ones resulted in another simple but highly successful Grumman naval fighter – the Cougar.

COUGAR

The Navy had commissioned three aerodynamically advanced fighters by the time Grumman's Panther entered active service – Vought's Cutlass in 1946, the Douglas Skyray in 1948 and McDonnell's Demon in 1949.

None of these three programmes made easy progress but the conventional straight-winged Panther with its reliable licence-made engine was there to pick up the slack. Encouraged by the success of the F9F-2, Grumman began working up plans to improve the Panther by giving it 35° swept back wings and swept tailplanes. The company proposed this change to the Navy and on March 2, 1951, was given a contract to build three examples under the designation XF9F-6.

With the rest of the Panther remaining largely unchanged, the work progressed quickly and the first prototype flew just over six months later on September 20. The surface area of the Panther's new wings was 40% greater than before and large trailing edge flaps were provided for increased low speed lift along with long leading edge slats. Flaperons replaced ailerons on top of each wing to control rolling movement.

Early flight testing showed that the modified Panther was now 50mph faster and low speed handling remained acceptable. Armament was still four 20mm cannon but only two underwing hardpoints could be fitted – below the wingroots, inboard of the point at which the swept wing section folded upwards. These tended to be used only for external fuel tanks if they were used at all, though the aircraft could carry anything up to a pair of 1000lb bombs on them if required.

GRUMMAN F9F-6 COUGAR

Grumman F9F-6 Cougar, A-218, VF-142, USS *Boxer* (CVA-21), 1955. The F9F-6 Cougar was a development of the earlier F9F-5 Panther introducing a new swept wing along with equipment and engine modifications.

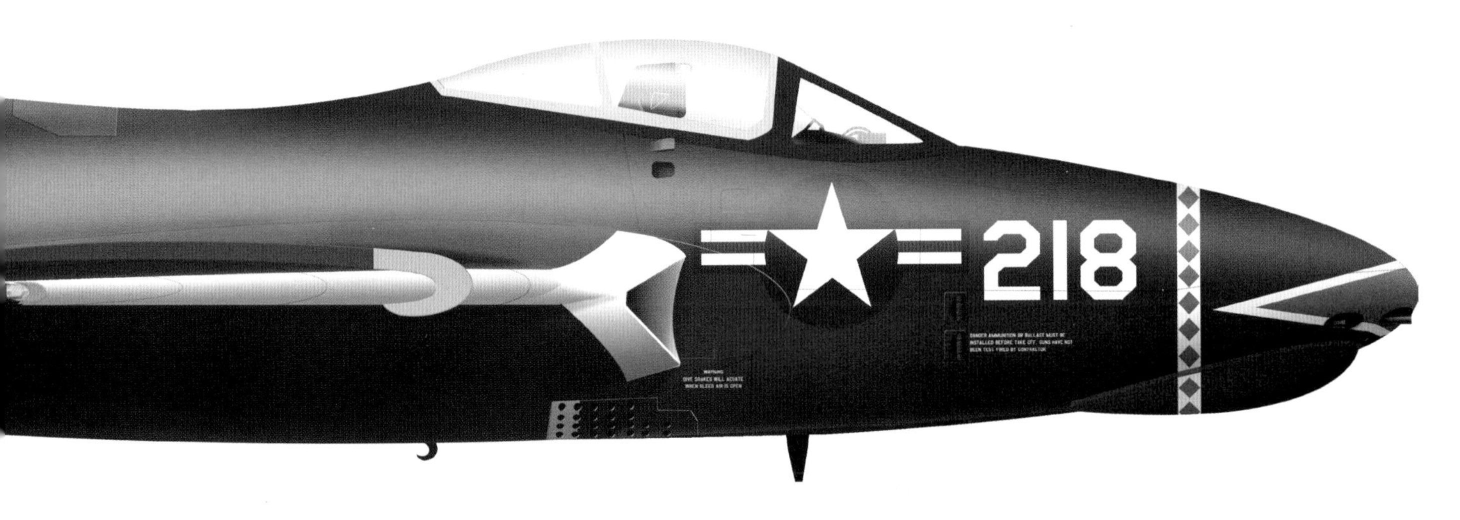

Grumman F9F Cougar

GRUMMAN F9F-7 COUGAR

Grumman F9F-7 Cougar, D-117, US Navy Reserve, NARTU, NAS Dallas, Texas, 1955. The Cougar equipped several units of the US Navy Reserve and continued to serve with the Reserve until the mid-1960s.

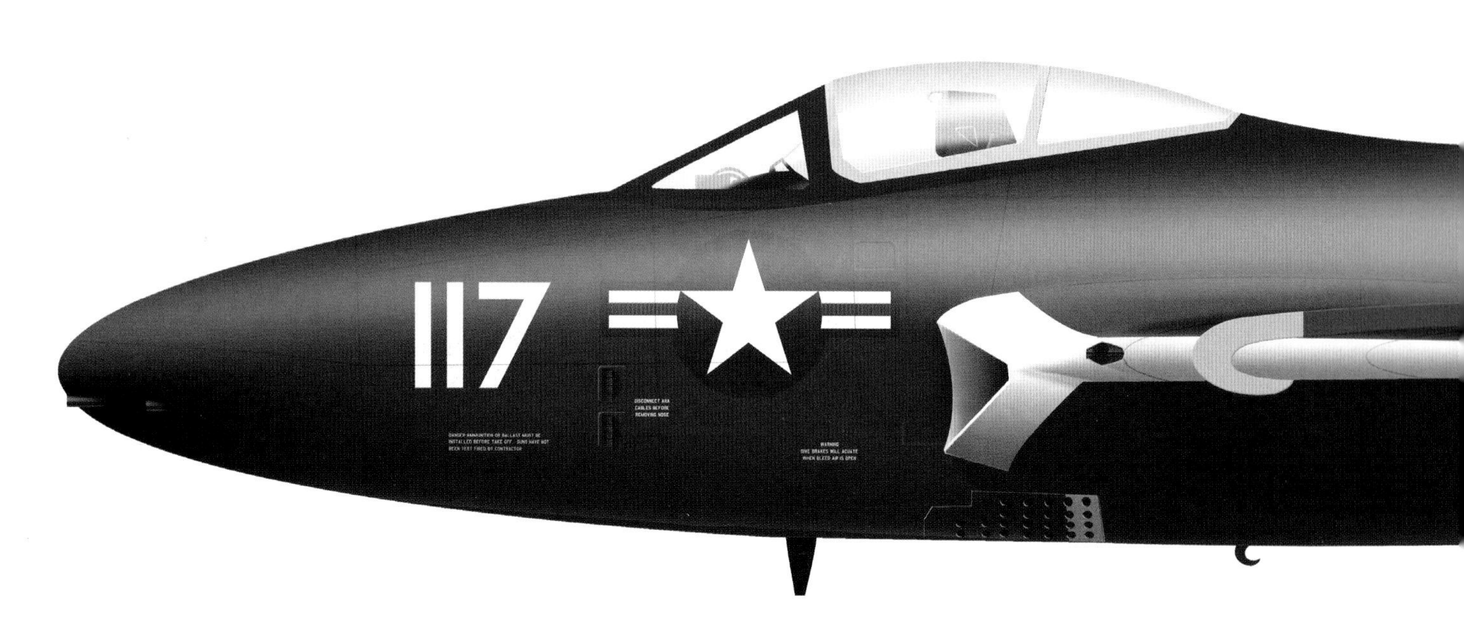

GRUMMAN F9F-6 COUGAR

Grumman F9F-6 Cougar, 7N-101, US Navy Reserve, NARTU, NAS Lincoln, Nebraska, 1957. US Navy Reserve was the final destination of many Cougars, following front line squadron service in the later part of the 1950s.

An order was placed for full production model aircraft and deliveries quickly commenced. Grumman decided that the Panther had been sufficiently modified to warrant a new name and called the swept wing version the Cougar.

It gradually became clear, however, that the additional performance was causing problems for the Cougar. Its tailplanes were fitted with conventional elevators which failed to work properly when the aircraft – designed for subsonic flight – was approaching supersonic speed during extreme manoeuvring. The solution, already introduced on the F-86A Sabre, was to give the Cougar all-moving tailplanes. The elevators remained but now each whole tailplane could be trimmed hydraulically. Another addition to the Cougar was wing fences for boundary layer control.

Deliveries of production model Cougars commenced in mid-1952 and the first fleet squadron to receive them was VF-32 'Fighting Swordsmen' towards the end of the year. VF-24 'Corsairs' followed shortly afterwards but was now too late for the type to see action in Korea. The Blue Angels display team had been due to trade their Panthers in for Cougars in 1953 but the elevator problem ▶

Grumman F9F Cougar

GRUMMAN F9F-8 COUGAR

Grumman F9F-8 Cougar, T-316, VF-13, USS *Bennington* (CVA 20), 1955. VF-13 operated the Cougar from 1954 to 1956.

meant this had to wait until the winter of 1954/55, with the first Cougar displays being flown during the 1955 season.

A total of 656 F9F-6s were delivered up to July 1954. Sixty of these were built as F9F-6P reconnaissance aircraft – the first such aircraft with swept wings to serve with the US Navy and the Marines. The F9F-6P had exactly the same nose section as that used on the F9F-5P Panther – since apart from the wing and tail surfaces they remained essentially the same aircraft.

Just as the F9F-3 had been an Allison-engined counterpart to the Pratt & Whitney-engined F9F-2, and likewise the F9F-4 had been the Allison-engined counterpart to the F9F-5, the F9F-6 also had its Allison-engined opposite number – the F9F-7. And for a third time exactly the same dynamic played out. While the F9F-6's Pratt & Whitney J48-P-8 (Rolls-Royce Tay) engine had proven both reliable and powerful (7250lb thrust), the F9F-7's Allison J33-A-16A was less powerful (5850lb thrust) and more troublesome. Once again a smaller number were built – just 168 – and once again most of these aircraft had their J33s removed and replaced with J48s.

The first F9F-7 flew in March 1953 and the production run lasted from the following month until June 1954. Some F9F-7s were actually delivered directly to Reserve units. A handful were converted into drones as F9F-7Ks or drone controllers as F9F-7Ds. Two were modified to take part in trials of a British-inspired carrier landing scheme known as the flexible deck project.

The basic concept behind this involved eliminating landing gear from the design of carrier aircraft. Since 33% of a typical carrier aircraft's weight was its retractable undercarriage, building aircraft without it would allow extra fuel and weapons to be carried. Taking off would be accomplished using a wheeled dolly that would be ejected on take-off and landing would involve bellying down on a soft rubbery deck surface combined with several layers of restraining cables.

GRUMMAN F9F-8 COUGAR

Grumman F9F-8 Cougar, O-310, VA-76, USS *Forrestal* (CVA-59), 1957. Cougars equipped this attack squadron from 1956 to 1959; VA-76 was deployed during the Suez crisis, operating in the Atlantic and ready to deploy to the Mediterranean if needed. F9F-8s could be equipped with a fixed nose-mounted in-flight refuelling probe.

Grumman F9F Cougar

The experimental deck surface, manufactured by Goodyear, was made of lubricated rubberised fabric 0.5in thick and was installed at the Naval Air Test Center, NAS Patuxent River, Maryland, in February 1955. A pair of F9F-7s had their Allison J33-A-16s swapped for the more powerful J48-P-8; their wings were modified and they had a 3in deep false underside fitted to their fuselages to provide a cushioning effect.

The outcome was a resounding failure.

Aircraft designed for flexible deck landings would be unable to carry under-wing stores – since anything not ditched before landing could easily cause an accident. Furthermore, aircraft that had landed could not be moved around without special equipment and a time-consuming process of recovery. And finally, unless a perfect landing was made the aircraft could easily slide or roll into a crash and it would be impossible to simply take off again and go around for another approach if that happened. Neither could the aircraft divert to a conventional landing field elsewhere in an emergency. Testing ended in August 1955 and the project was cancelled.

The next step in the Cougar's evolution, the F9F-8, had both redesigned wings and a redesigned fuselage. The latter was stretched by 8in to provide further space for the internal fuel tanks – an extra 140 gallons – and the former saw its leading edge slats deleted while the chord of the outer wing was increased. The wingroot fillets were also enlarged, now going all the way back to the end of the fuselage.

The new variant received an uprated engine too. The J48-P-8A could provide a whopping 8500lb of thrust with water injection – a far cry from the 5000lb produced by the J42-P-6 fitted to the original F9F-2 prototypes – but much of the gain over the J48-P-8 was swallowed up by the aircraft's additional bulk.

The F9F-8 was also fitted with a second hardpoint under each wing, on the swept section, which could carry rocket pods, bombs up to 500lb or Sidewinder missiles.

The first F9F-8 flew on December 18, 1953, and it only took just over two months to get the full production line up and rolling. The first deliveries of full series F9F-8s began on February 29, 1954, continuing up to March 22, 1957, when production ended with a total of 601 having been built.

Such was the success of the F9F-8 Cougar that it would eventually be flown by 30 Navy and Marine squadrons – both fighter and attack units – and it was the most common aircraft in Navy service for several years.

There were several variants based on the standard F9F-8: the F9F-8B was equipped to launch tactical nuclear weapons using the radical bomb toss manoeuvre known as LABS – Low

GRUMMAN F9F-8B (AF-9J) COUGAR

Grumman F9F-8B (AF-9J) Cougar, 3H-358, VT-23, NAS Kingsville, Texas, 1967. Training squadrons used not only the two-seat variants of the Cougar, but also single-seaters until the late 1960s; after 1962, the F9F-8B was redesignated AF-9J.

GRUMMAN F9F-8B COUGAR

Grumman F9F-8B Cougar, 072, NOTS, NAS China Lake, California, 1956. Cougars were used for weapons and armament testing at the Naval Ordnance Test Station; this aircraft carries AIM-9B Sidewinder AAMs.

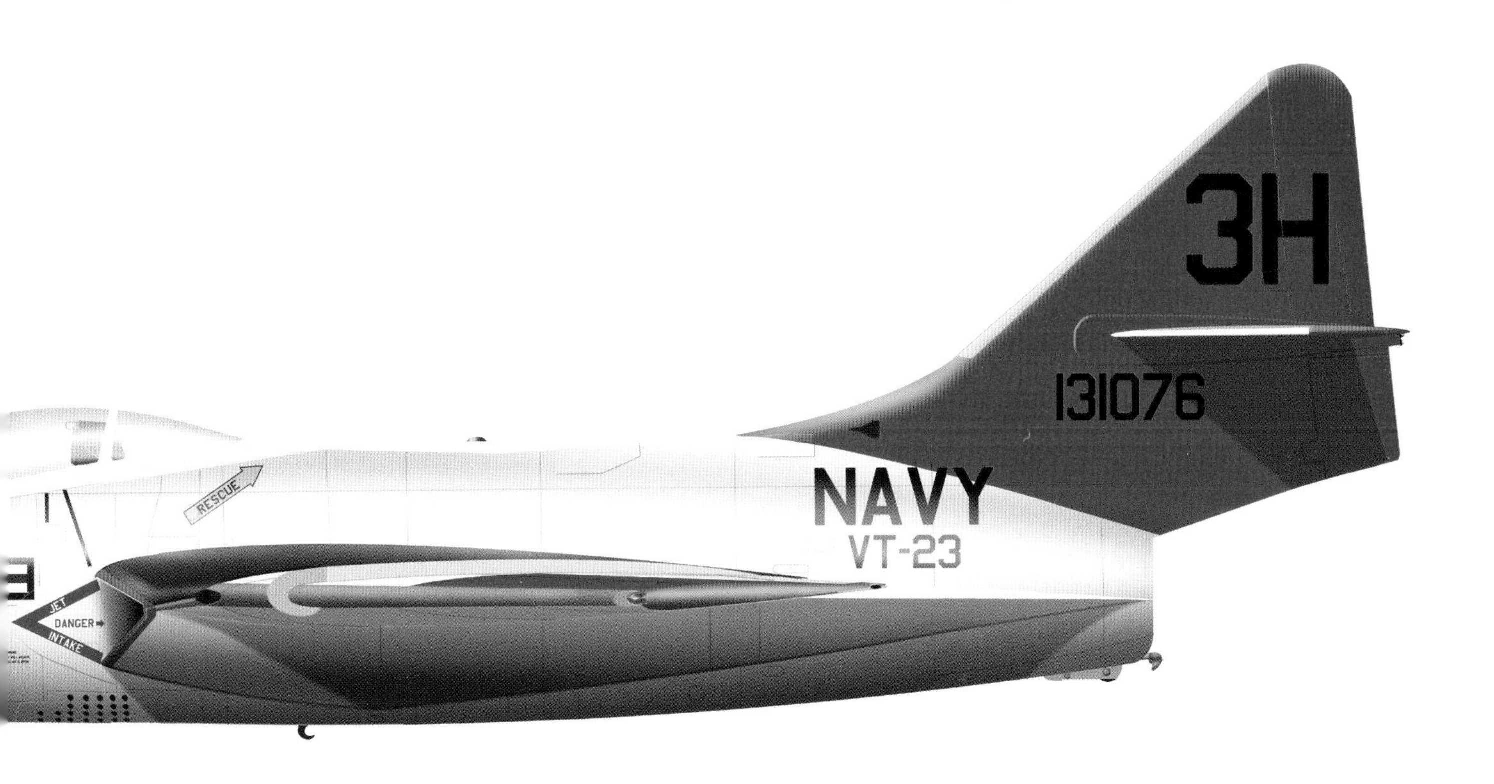

GRUMMAN F9F-8B COUGAR

Grumman F9F-8B Cougar, ND-103, VF-94, NAS Moffett Field, California, 1957. VF-94 flew the Cougar from 1955 to 1957.

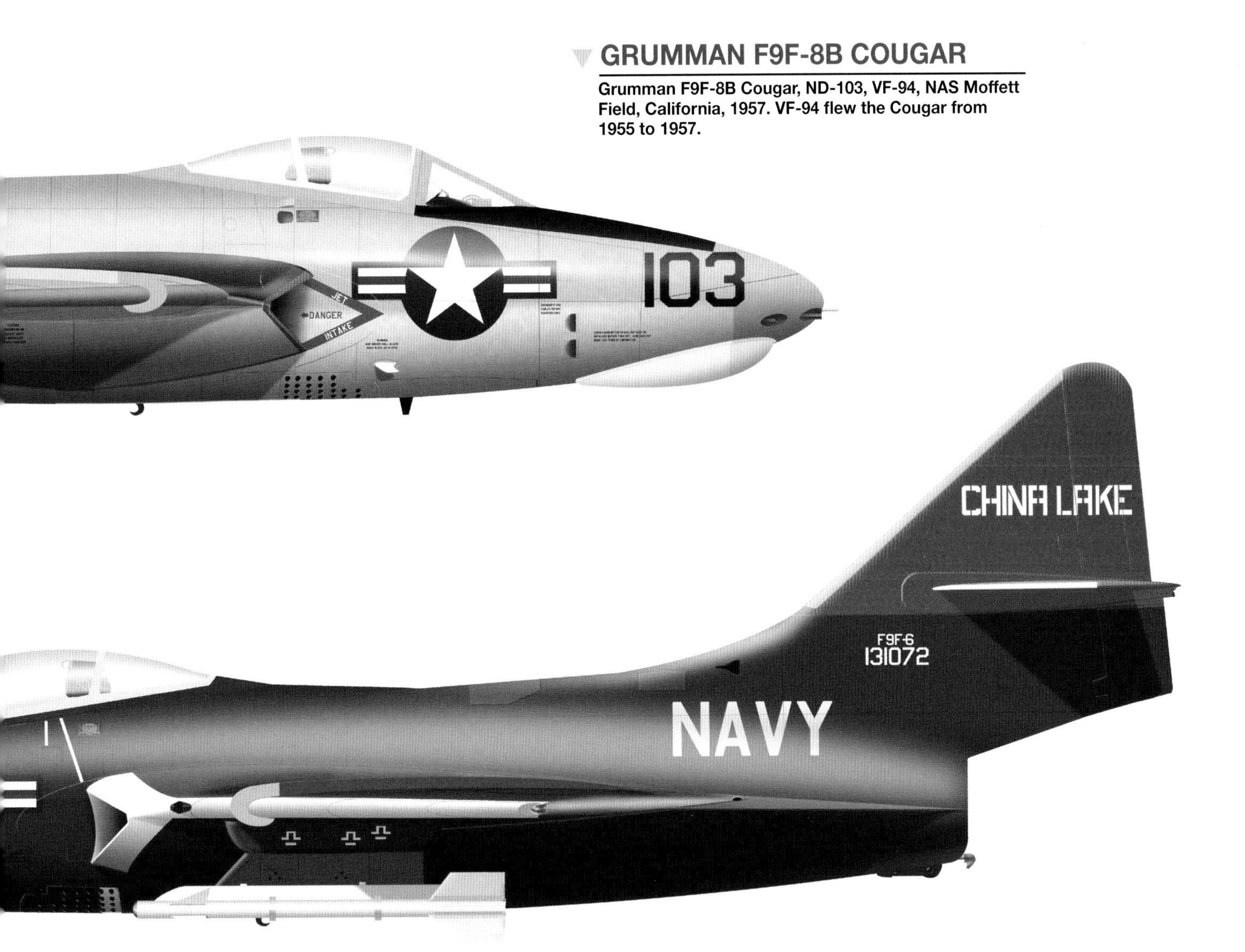

Grumman F9F Cougar

GRUMMAN F9F-8T COUGAR

Grumman F9F-8T Cougar, 448, NPF, NAS El Centro, California, 1961. The Naval Parachute Facility used this two-seat Cougar for ejection seat testing; it has the rear canopy removed as a result.

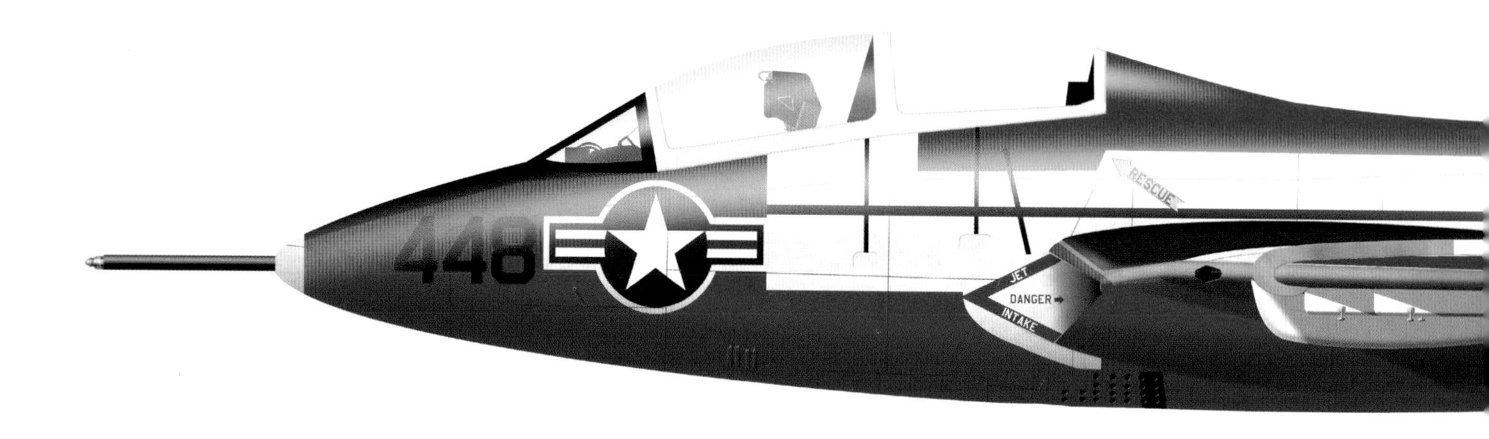

GRUMMAN F9F-8T COUGAR (TF-9J)

Grumman F9F-8T Cougar (TF-9J), 0, Blue Angels, 1965. The Blue Angels flew the Cougar from 1955 to 1957; a two-seat aircraft remained with the team until the late 1960s and was used for press and VIP flights.

GRUMMAN F9F-8T (TF-9J) COUGAR

Grumman F9F-8T (TF-9J) Cougar, 2F-353, VT-4, NAS Pensacola, Florida, 1974. The two-seat variant of the Cougar enjoyed a prolonged service with the US Navy, being used by training units until the mid-1970s; VT-4 was the last unit to retire the Cougar in 1974.

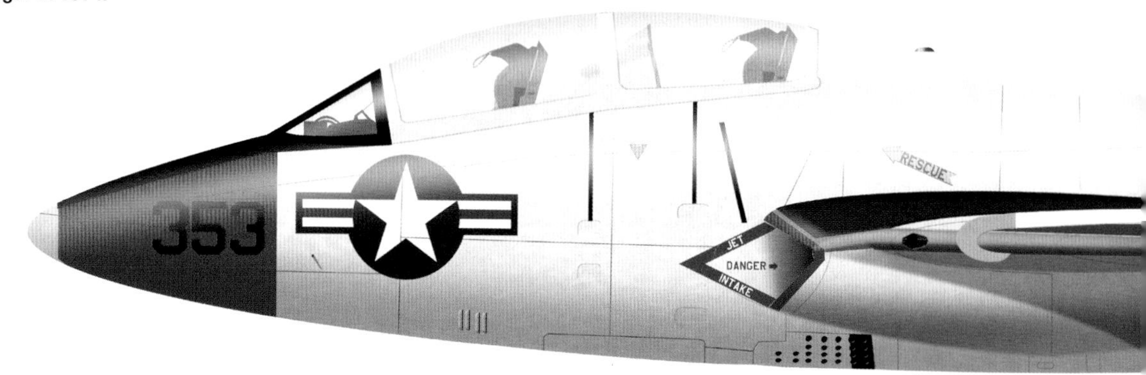

Altitude Bombing System. It was fitted with an Aero 22A rack beneath its right wing capable of carrying a load of up to 2000lb and internal electronics which allowed the pilot to make the difficult LABS delivery effectively.

The F9F-8P reconnaissance variant differed significantly from earlier Panther and Cougar 'P' types in having a much longer nose. This lengthy snout had seven apertures and accommodation to suit a wide variety of different cameras. A total of 110 F9F-8Ps were produced and served up to the end of 1960 when they were replaced by newer types.

The F9F-8T trainer variant required another fuselage extension to accommodate a second cockpit behind the original one. Another 34in section was added and internal fuel tank capacity had to be reduced so that the rear position could extend further back into the fuselage. Two of the four cannon were also removed.

By September 1962 there were still numerous Cougars serving with the Navy and they received new designations along with all other 1950s aircraft. The F9F-6 became the F-9F, F9F-6P became RF-9F, F9F-7 became F-9H, F9F-8 became F-9J, F9F-8B became AF-9J and F9F-8T became TF-9J.

Under this new name, several two-seat Cougars were used in Vietnam for forward air control duties. The back-seater would be tasked with noting enemy positions and then directing both ground forces and air support onto them.

As with other early fighters, many F-9J Cougars ended up as drones and drone controllers. These became QF-9Js. The Cougar remained in service as an advanced trainer into the early 1970s, demonstrating the adaptability and fundamental strength of the original Panther design – drawn up more than two decades earlier. ●

VOUGHT F7U CUTLASS

The Cutlass was a leap into the unknown, embodying numerous advanced features. Unfortunately, this was still the 1950s and the aircraft's primitive systems ensured that its performance could not live up to the promise of its exotic looks.

1951-1959

VOUGHT F7U-1 CUTLASS

Vought F7U-1 Cutlass, 7, Blue Angels, 1953. During the 1953 season, the Blue Angels acquired two F7U-1 aircraft, but abandoned plans to replace their existing aircraft with it.

When the Second World War ended the US was able to take advantage of copious research carried out by German engineers and scientists on swept wings and tailless aircraft design. While the Americans already had some experience in these areas, hundreds of thousands of captured German documents supplied invaluable additional data.

Tailless design meant deleting an aircraft's horizontal tail surfaces – the parts thought to cause the most problems during transonic flight – and swept-back wings had been shown to offer significant performance advantages.

With several 'safe' and conventional designs already in progress, including the Fury, the Banshee and Vought's own XF6U-1 Pirate, the Navy felt it could take a gamble on a new and entirely unconventional design offered by Vought. Three prototypes were ordered under the designation XF7U-1 on June 6, 1946, and Vought stuck with its swashbuckling naming policy by applying the moniker 'Cutlass'.

The new aircraft's slender fuselage housed a pair of Westinghouse J34-WE-32 engines side by side at the rear fed by wing root intakes – but had no fins or tailplanes on it. Instead, the aircraft's 38° swept-back wings each had a fin attached part way along its trailing edge.

The tricycle undercarriage included an unusually long nosewheel leg which put the pilot 14ft up in the air even before take-off. An ejection seat was fitted but this was only effective above 1500ft.

Armament was four 20mm cannon under the cockpit in the nose and there was no option to carry weapons or fuel tanks externally at this stage. The Navy was so committed to the Cutlass that it ordered 20 production F7U-1s on July 29, 1948, even before the type's first flight.

Vought test pilot Joel Robert 'Bob' Baker took the first prototype up for its aerial debut on September 29, 1948, and testing progressed well with the second and third prototypes following in sequence. Vought proposed a second production model – the F7U-2, powered by Westinghouse J34-WE-42 engines – on November 1, 1948. The Navy came through with an order for 88 of these.

The second prototype was lost during a test flight on March 14, 1949, over Chesapeake Bay, Maryland. Test pilot William Millar vanished into a bank of cloud and pieces of his Cutlass washed up on the shore a month later. His body was never recovered.

VOUGHT F7U-1 CUTLASS

Vought F7U-1 Cutlass, TT-415, NATC, USS *Midway* (CVB-41), 1951. This aircraft was the first Cutlass used for carrier tests.

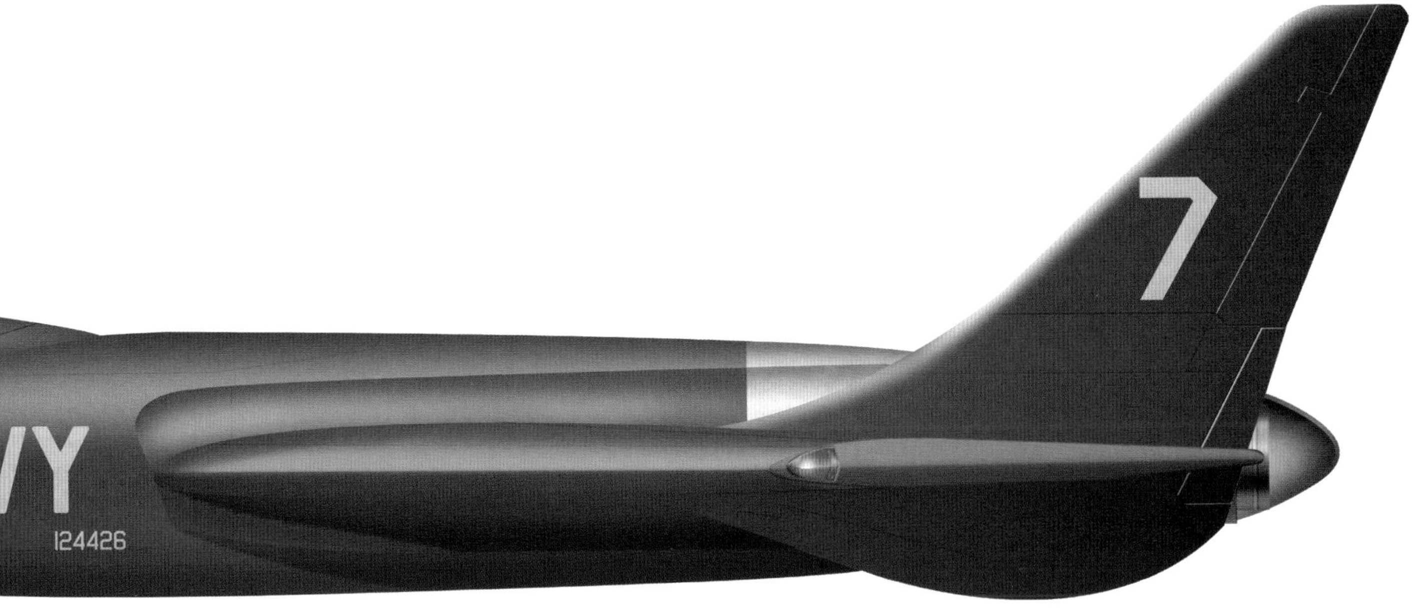

Vought F7U Cutlass

Almost a year to the day after its maiden flight, the first prototype was also destroyed. Test pilot Paul Thayer overcorrected for a crosswind on take-off while carrying a heavy asymmetrical load on September 28, 1949. The aircraft's port wing dipped until it hit the ground to the left of the runway. After spinning around 100°, it slid 750ft across the ground, its fuselage breaking away forward of the engine intakes. Thayer escaped with cuts and bruises.

The first production F7U-1 flew on March 1, 1950. It had been hoped that the F7U-1 would be powered by Westinghouse's afterburning J46 but this engine was unavailable so the J34 continued to be used – along with 1245lb of ballast to simulate the bigger, heavier powerplant.

Only 14 of the 20 F7U-1s ordered were actually built and two of them crashed during test flights at Vought's factory. The Blue Angels demonstration team briefly flew two more during solo displays but this activity was soon curtailed due to a lack of spare parts. The remaining 10 were used for carrier take-off and landing tests from June 24, 1950, to August 14, 1951. The third and last surviving prototype was destroyed after catching fire mid-air during an air show on July 7, 1950.

During testing of the F7U-1 it became clear that Westinghouse was never going to come through with the J34-WE-42 and the F7U-2 was cancelled. However, Vought had already pitched a much more substantial upgrade as the F7U-3.

The F7U-3 retained the basic layout of the F7U-1 but was otherwise a completely new aircraft. The whole airframe was enlarged and strengthened. Wingspan was increased by a foot, also increasing wing area, and the landing gear was strengthened – particularly the nosewheel. The cockpit was raised even higher and redesigned to provide better visibility over the nose during take-off and landing.

Vought added more than 100 new access panels too, making maintenance of the aircraft considerably easier, and the rear fuselage was redesigned so that the engines could be slid out rearwards. In the F7U-1 they had had to be lowered out of the aircraft.

The four 20mm cannon were removed from their original nose mountings and repositioned much further back, two above each engine intake. The F7U-3 could also carry a substantial payload – being capable of a catapult launch carrying a full internal fuel load plus 5500lb of external stores.

Engines continued to be a problem, however. The first 16 F7U-3s had to be powered by Allison J35-A-29s because the afterburning J46-WE-8A remained unavailable. When the J46-WE-8A did finally arrive its output was well shy of the figures Westinghouse had promised – 3620lb of thrust per engine, rising to 5725lb with afterburner. Even so, a ▶

VOUGHT F7U-3 CUTLASS

Vought F7U-3 Cutlass, 406, VF-83, USS *Intrepid* (CV-11), 1954. VF-83 operated the Cutlass from 1954 to 1957.

VOUGHT F7U-3 CUTLASS

Vought F7U-3 Cutlass, 10, VA-12, NAS Cecil Field, Florida, 1957. The Cutlass was also deployed by attack squadrons and could be equipped with a ventral pack containing 2.75in folding fin rockets.

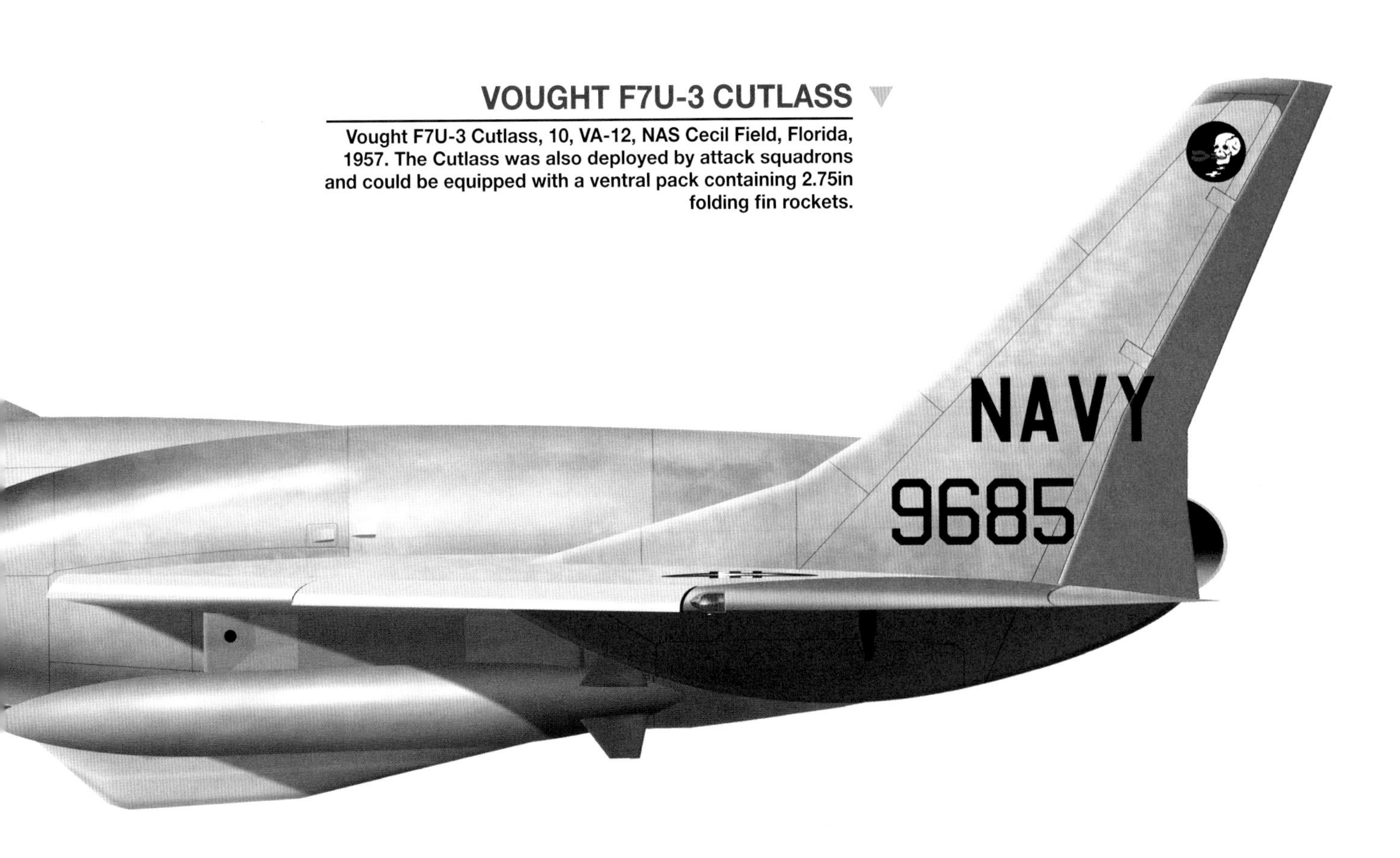

Vought F7U Cutlass

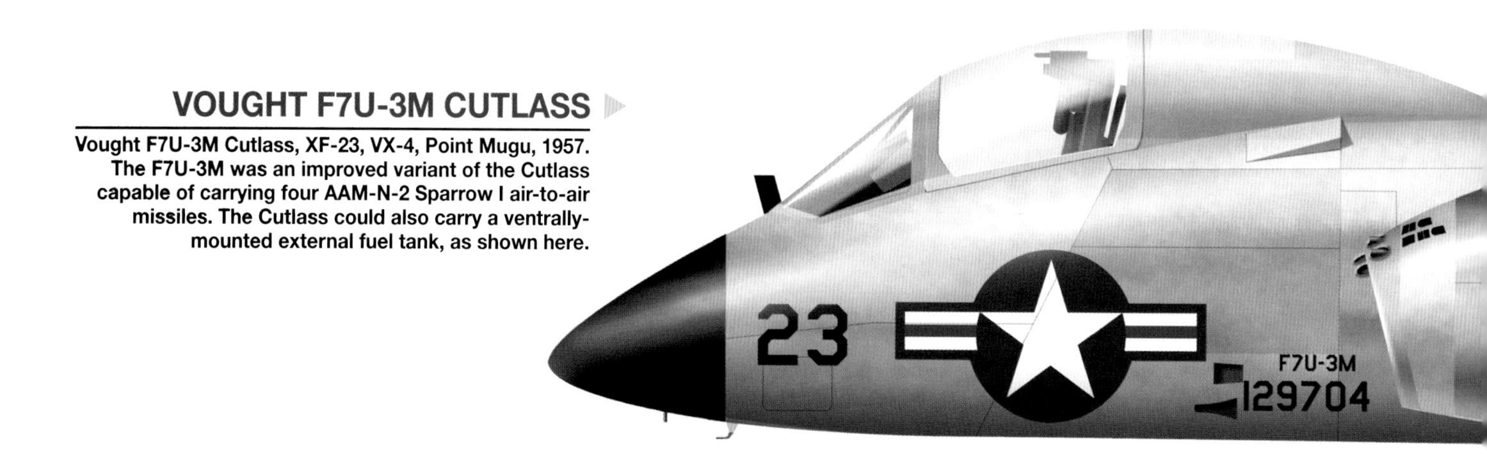

VOUGHT F7U-3M CUTLASS

Vought F7U-3M Cutlass, XF-23, VX-4, Point Mugu, 1957. The F7U-3M was an improved variant of the Cutlass capable of carrying four AAM-N-2 Sparrow I air-to-air missiles. The Cutlass could also carry a ventrally-mounted external fuel tank, as shown here.

J46-engined Cutlass outperformed both a Grumman F9F-6 Cougar and North American FJ-2 Fury in mock combat.

The Navy ordered 180 F7U-3s and they were beginning to reach units by June 1954, including VF-81 and VF-83 operating in the Atlantic and VF-122 and VF-124 in the Pacific. A total of 13 squadrons would operate the type.

Ninety-eight of the 180 F7U-3s – more than half – were built as F7U-3Ms, with the ability to carry two Sparrow air-to-air missiles under each wing. They were fitted with AN/APQ-51 radar guidance systems and also could be fitted with an inflight refuelling probe.

A further 12 F7U-3s were converted into F7U-3Ps for photo reconnaissance, with cameras being installed in a nose 25in longer than the original. Flare ejectors with 40 cartridges were fitted behind the cockpit so that targets could be illuminated during night photography. These aircraft were never used in operational service but were used extensively for testing and evaluation.

VOUGHT F7U-3 CUTLASS

Vought F7U-3 Cutlass, D-415, VF-124, USS *Hancock* (CVA-19), 1956. Later aircraft were equipped with a nose-mounted in-flight refuelling probe; this feature was retrofitted to some earlier aircraft.

While the F7U-3 Cutlass was not as 'gutless' and its unfortunate nickname – 'Gutless Cutlass' – might suggest, it was reportedly a difficult aircraft to fly and its systems let it down. For example, its hydraulics system operated at 3000psi, which was twice that of other Navy jets. As a result, it constantly leaked and lost pressure, resulting in unresponsive controls in the cockpit.

In addition, under certain conditions the afterburners of the J46 could use up all the fuel in the central transfer tank before more could be pumped in from the full wing tanks, resulting in flame-out on take-off. Inflight engine fires, as previously mentioned, were all too frequent.

The heavy Cutlass was also extremely tricky to recover if it got into a spin. Taken together, these characteristics resulted in a high number of life-threatening incidents and earned it another nickname, the 'Ensign Eliminator'. A total of 21 Navy pilots died from 78 accidents and a quarter of all Cutlasses were destroyed as a result of those accidents.●

MCDONNELL F3H DEMON

McDonnell's successor to the Banshee, the Demon, was perhaps the fighter worst affected by engine manufacturer Westinghouse's inability to develop a truly satisfactory jet. Eventually, with an Allison powerplant, it became a successful all-weather fighter.

MCDONNELL F3H-2N DEMON

McDonnell F3H-2N Demon, D-214, VF-124, USS *Lexington* (CVA-16), 1957. VF-124 was the first Pacific Coast squadron to receive the Demon, in 1956. It made its debut cruise aboard the USS *Lexington* in 1957.

1956-1964

The US Navy issued a requirement for a fast-climbing high-altitude fleet defence interceptor in 1947 and designs were tendered by both McDonnell and Douglas. Where the latter's proposed aircraft was tailless, like the Cutlass, McDonnell's submission had a conventional layout – albeit with sharply swept wings and tail surfaces.

On September 30, 1949, the Navy ordered two prototypes of each and all four single-jet aircraft were to be powered by Westinghouse's new J40 engine, which the company said would produce 11,600lb of thrust with the aid of an afterburner.

McDonnell's design was the XF3H-1 and the company named it 'Demon'. Part way through development, in March 1951, the Navy decided that it was needed to replace the Banshee as all-weather interceptor. One hundred and fifty production model F3H-1Ns were ordered and it was specified that these should be equipped with a nose-mounted AN/APQ-50 radar.

The first prototype made its flight debut on August 7, 1951. However, the first version of the J40 fitted, the non-afterburning J40-WE-6, caused real problems right from the outset. It generated just 7500lb of thrust and suffered from severe technical and reliability issues.

Westinghouse quickly replaced this with the J40-WE-8, which had an afterburner installed, but this version also had substantial flaws and reliability issues. To make matters worse, even when it worked it still only produced 10,900lb of thrust.

When the engine did work, the two XF3H-1s had more than satisfactory performance. But more often than not the aircraft were grounded by technical issues. Eventually the second prototype was made useable enough for carrier trials, commencing them in August 1953 after lengthy delays. Seven months later, however, the first prototype was destroyed by an on-board fuel explosion later linked to a problem with the J40. The second prototype was immediately grounded and never flew again.

The F3H-1Ns were to be powered by Westinghouse's J40-WE-22, the production model version of the J40-WE-8, but this was bigger than its predecessor which meant that the ▶

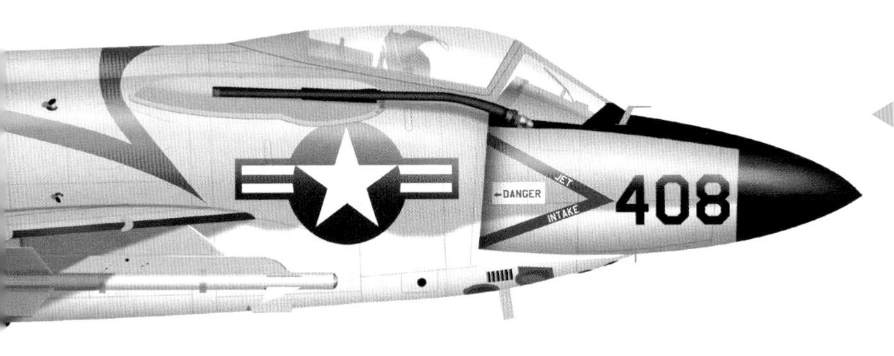

MCDONNELL F3H-2N DEMON

McDonnell F3H-2N Demon, NH-408, VF-114, USS *Shangri-La* (CVA-38), 1958. Although the Demon could carry external fuel tanks, it was found that this did not increase its range (and could even reduced it due to the added drag and weight of the tank); the addition of an in-flight refuelling probe was effective, however.

McDonnell F3H Demon

MCDONNELL F3H-2 DEMON

McDonnell F3H-2 Demon, AJ-107, VF-82, USS *Ranger* (CVA-61), 1958. The F3H-2 was designed with a strike capability, able to carry air-to-ground ordnance in the six under-wing and the two fuselage pylons. It retained the capacity to carry air-to-air missiles as well.

MCDONNELL F3H-2N DEMON (F-3C)

McDonnell F3H-2N Demon (F-3C), 4, NATF, Lakehurst, New Jersey, 1963. The Naval Air Test Facility used this Demon for several test programmes. Under the new Tri-Service aircraft designation system the Demon became the F-3.

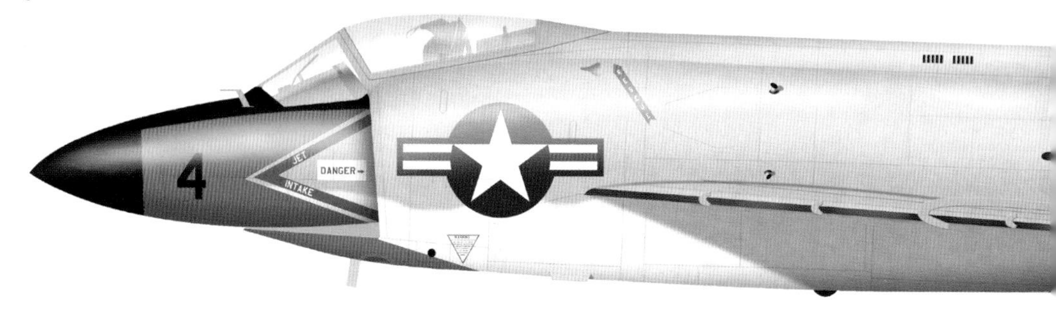

MCDONNELL F3H-2M DEMON

McDonnell F3H-2M Demon, NH-209, VF-112, NAS Miramar, California, 1958. The Demon could carry four AIM-7 Sparrows; initially destined to carry the Sparrow I, it was later decided to fit the improved Sparrow III. Carrying AAMs frequently led to the removal of two of the four guns.

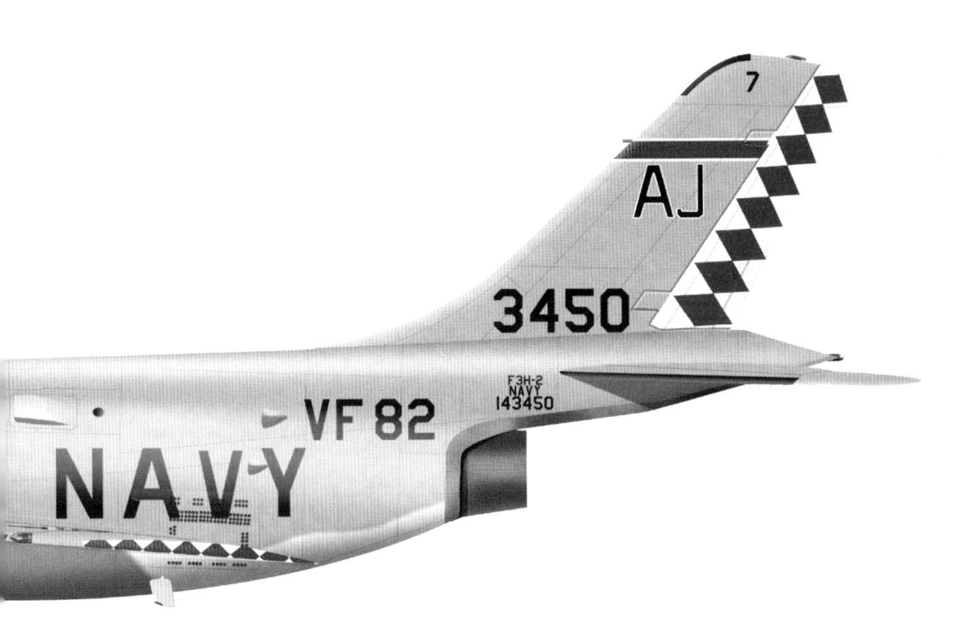

MCDONNELL F3H-2 DEMON (F-3B)

McDonnell F3H-2 Demon (F-3B), AK-104, VF-13, USS *Shangri-La* (CVA-38), 1963. The F3H-2 variant introduced the short beaver tail; this features was also retrofitted to earlier variants. The ability to carry both the AIM-9 sidewinder and the AIM-7C Sparrow III missiles allowed the Demon to improve its fleet defence mission.

Demon's whole fuselage had to be enlarged to accommodate it. Armament of four 20mm cannon was also added. The first F3H-1N flew on Christmas Eve, 1953, but deliveries had barely commenced when a series of fatal crashes almost derailed the entire programme.

Six aircraft were lost in eight crashes with four pilots being killed. Eventually most of these accidents were linked to defects in the J40. Just 56 of the 150 F3H-1Ns ordered had been built. None of the surviving aircraft entered operational service – in fact very few of them ever flew. It was a disaster for the Navy, for McDonnell and most of all for Westinghouse. With all the evidence pointing to the engine manufacturer as the source of the Demon's problems, the Navy agreed to continue the programme with a different powerplant.

Allison's J71-A-2 was chosen to replace the J40 and McDonnell took the opportunity to carry out a number of further alterations to the Demon's design. The wings were made deeper at their root, resulting in greater wing area without increased wingspan, and the AN/APG-51A radar replaced the AN/APQ-50. The Demon was also provided with the equipment required to carry and launch up to four Sidewinder missiles. Alternatively, up to 4000lb of external stores could be mounted on four underwing pylons.

This revised aircraft entered service as the F3H-2N in March 1956 and 142 were built. Some reliability issued were experienced with the J71 but it was nevertheless a great deal better than the J40.

In common with other Navy jet fighters of the period, a dedicated 'missile' version of the Demon was built under the designation F3H-2M to carry the beam-riding Sparrow I. A total of 79 were built but F3H-2M pilots had exactly the same problems as everyone else who tried to use the Sparrow I – it was woefully unreliable and thoroughly impractical for use in combat. The F3H-2M therefore only lasted around three years before being retired.

While the F3H-1N had been a disaster, the F3H-2N had been a passable success and the F3H-2M had been another failure, the Demon hit its stride at last with the F3H-2 – which finally dropped both the 'N' and the 'M' even though it had an AN/APQ-51B search radar for all-weather operations and the ability to launch the semi-active radar-homing Sparrow III missile. It was even fitted with an improved J71 to cure that engine's reliability problems.

A total of 239 F3H-2s were built and some F3H-2Ns were retrofitted to F3H-2 standard. When the designation system was standardised in 1962, the F3H-1N became the F-3A, the F3H-2 became the F-3B, the F3H-2N became the F-3C and the F3H-2M became the MF-3.●

DOUGLAS F4D-1 SKYRAY

1956-1965

Developed in parallel to McDonnell's Demon, the tailless Douglas Skyray ended up with an excellent engine following the failure of the Westinghouse J40 which made it a capable fighter/interceptor.

DOUGLAS F4D-1 SKYRAY

Douglas F4D-1 Skyray, PA-32, VFAW-3, Norton Air Force Base, California, 1961. The VFAW-3 was integrated into the 27th Air Division of the USAF (part of NORAD) and assigned a continental air defence role.

Having studied test data captured from Germany in the wake of the Second World War, Douglas saw the benefits of both swept wings and a tailless layout. The design it tendered to meet the Navy's 1947 requirement for a fast-climbing high-altitude fighter therefore included both features.

Like McDonnell's tender for the same contract, Douglas's aircraft was to be powered by the initially very promising Westinghouse J40. It was given the designation XF4D-1 and two prototypes were ordered on September 30, 1949 – the same day that McDonnell received its contract for two Demon prototypes.

The XF4D-1 was given the name 'Skyray' since in planform it looked rather like a manta ray and because Douglas had decided to establish a convention of using the 'Sky' prefix for all of its aircraft names. The aircraft's single engine was centrally mounted within a capacious fuselage and the large wings on either side extended from wingroot intakes just aft of the cockpit all the way back to the tail end. A single swept fin was mounted above the engine exhaust.

The aircraft had a simple tricycle undercarriage and was to be armed with four 20mm cannon. These were positioned in the underside of the aircraft's wings about halfway along, just inboard of the point at which the wings folded. Provision was also made to carry external stores.

Development progressed quickly and the first XF4D-1 prototype was approaching completion before even Westinghouse had got the non-afterburning J40-WE-6 ready, so the aircraft was fitted with an Allison J35-A17 instead – as was the second prototype. Where the J40, with afterburner, was expected to produce 11,600lb of thrust, the J35 was capable of just 4900lb. Despite this handicap, testing went ahead and allowed early problems with the aircraft's control surfaces to be ironed out.

A pair of J40-WE-6s were eventually delivered to Douglas in mid-1952 but, like McDonnell, the company immediately encountered problems with them when they were installed in the two XF4D-1 ▶

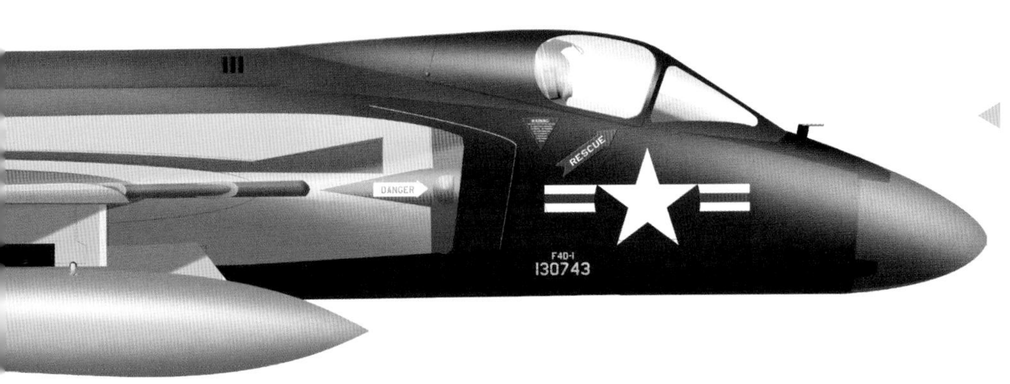

DOUGLAS F4D-1 SKYRAY

Douglas F4D-1 Skyray, 0743 NATC, NAS Patuxent River, Maryland, 1960. The Naval Air Test Center used this Skyray as a target tow aircraft; it carries the TDU-10B aerial target.

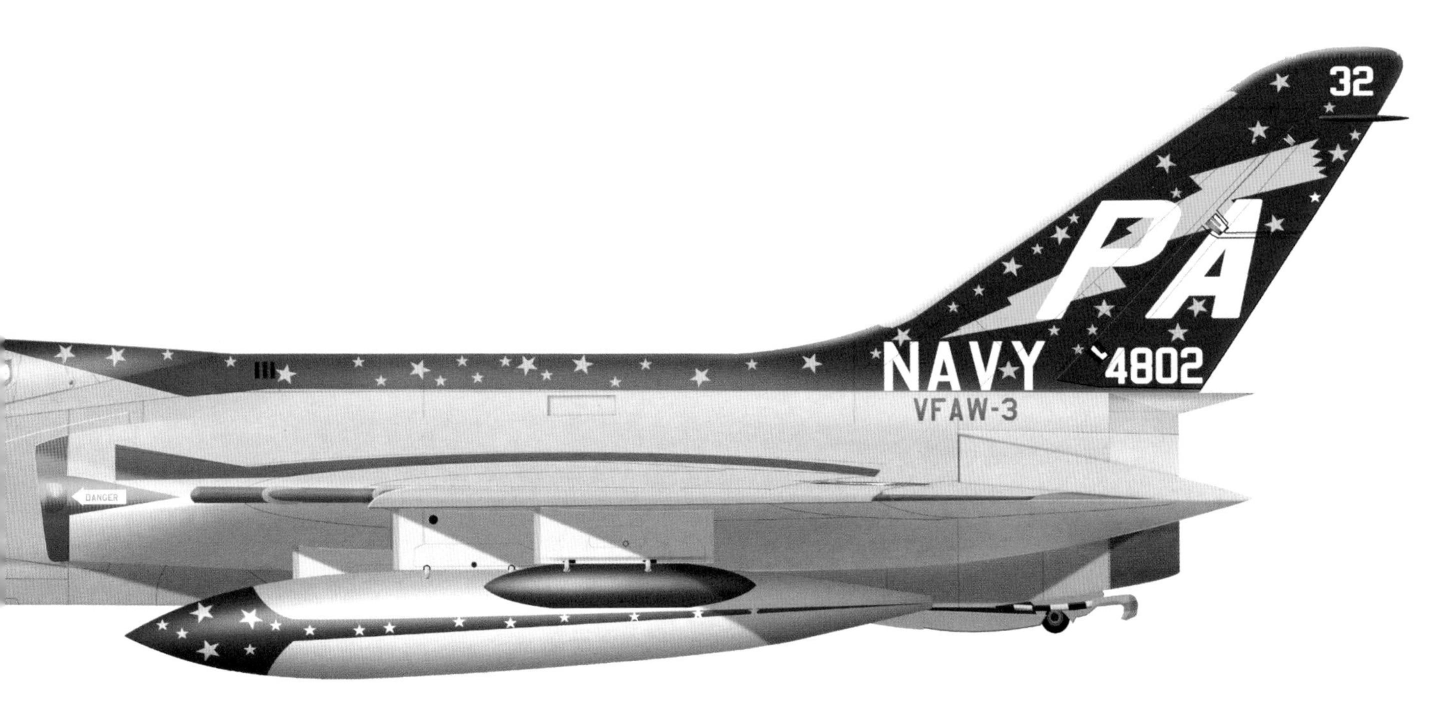

Douglas F4D-1 Skyray

prototypes. Their reliability was poor, they suffered from technical issues and their power output was way below what had been promised.

A year of development time was effectively wasted thanks to these engines and the much-anticipated J40-WE-8s were not delivered until mid-1953. With 10,900lb of thrust using the afterburner, this engine again fell slightly short of Westinghouse's promise but when it worked it offered good performance.

The second XF4D-1 Skyray, fitted with its J40-WE-8 and stripped of absolutely all non-essential equipment, was used to set a new absolute air speed record of 752.9mph on October 3, 1953, with Lieutenant Commander James B. Verdin at the controls. This made the Skyray the first US Navy aircraft to go supersonic.

Verdin had taken the title from British test pilot Mike Lithgow, who had only achieved his record of 735.7mph a week earlier on September 26 while flying a Supermarine Swift F.4. Verdin's own record fell at the end of the month when USAF chief test pilot Frank K. Everest flew an F-100 at 755.1mph. Verdin was killed just 15 months later when he ejected from a YA4D-1 Skyhawk prototype at 30,000ft but his parachute failed to open.

Again like McDonnell, Douglas found that the J40 was simply too unreliable and potentially dangerous to use. A replacement was needed but unlike McDonnell's Demon the Skyray's fuselage was large enough to accommodate a substantially bigger engine. The choice of powerplant for the production F4D-1s was narrowed down to Pratt & Whitney's excellent J57-P-2 – which produced 10,200lb of thrust even without its afterburner and 14,500lb with it activated. This was then changed to the J57-P-8 which managed 8700lb dry and an eye-watering (for the time) 16,000lb with afterburner.

The first production F4D-1 flew on June 5, 1954, with enlarged engine intakes but otherwise externally similar to the prototypes. In addition to their four cannon, the production aircraft could carry Sidewinders, rocket pods or external fuel tanks on six underwing pylons and one centre fuselage hardpoint. The aircraft's nose housed an AN/APQ-50 radar system.

Nearly two years of testing took place before the first deliveries were finally made to Composite Squadron VC-3 on April 16, 1956. A total of 420 Skyrays were built and there were no variants other than the standard F4D-1. An F4D-2 was initially planned but this was eventually built as the F5D-1 Skylancer – which, though it looked similar to the F4D-1 was a completely new aircraft. The Skylancer was not ordered by the Navy.

In 1962, surviving Skyrays received the designation F-6A. Under this number it continued to fly with Reserve units into 1965 before it was finally phased out.●

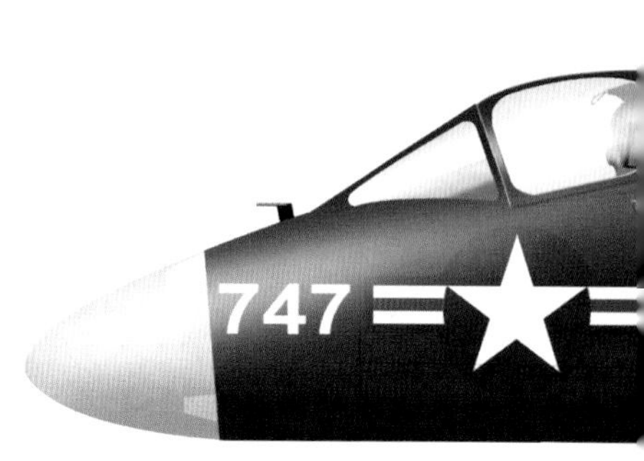

DOUGLAS F4D-1 SKYRAY

Douglas F4D-1 Skyray, 0747, Naval Ordnance Test Station, China Lake, California, 1960. NOTS used the Skyray as the launch vehicle for the Pilot project (known as NOTSNIK), designed to deliver satellites into space; the rocket was the NOTS-EV-1.

DOUGLAS F4D-1 SKYRAY

Douglas F4D-1 Skyray, AF-109, VF-74, USS *Intrepid* (CVA-11), 1959. VF-74 operated from the USS *Intrepid* and later the USS *Franklin D. Roosevelt* aircraft carriers, before becoming the first squadron to deploy with the new F-4 Phantom II.

DOUGLAS F4D-1 SKYRAY

Douglas F4D-1 Skyray, AF-103, VF-162, USS *Intrepid* (CVA-11), 1961. The Skyray did not have its own in-flight refuelling probe, instead using a probe attached to the port external fuel tank.

DOUGLAS F4D-1 SKYRAY

Douglas F4D-1 Skyray, NP-305, VF-213, USS *Lexington* (CVA-16), 1958. VF-213 deployed their Skyrays during the 1958 Taiwan Strait crisis; this aircraft carries AIM-9B sidewinder AAMs.

GRUMMAN F11F-1 TIGER

1956-1969

The final development of the successful Panther/Cougar line, the Tiger was a lightweight fighter at a time when greater range, heavier firepower and increased external loads were becoming ever more desirable. It suffered reliability issues and its front line career was brief.

GRUMMAN F11F-1 TIGER

Grumman F11F-1 Tiger, Blue Angels, 1957. The Blue Angels used, in succession, both short and long nose variants of the Tiger, from 1957 to 1968, long surpassing the time the Tiger was operated by embarked squadrons. This aircraft has a tube on the side of the fuselage, allowing the injection of oil into the exhaust to generate smoke as part of the display routine.

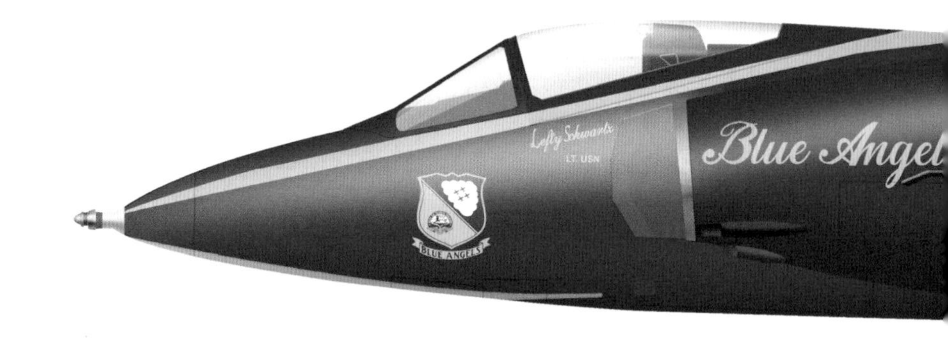

With the Cougar, Grumman decided that they had a fighter with the potential to go supersonic. The company therefore embarked on a privately-funded development programme which would see the original Panther/Cougar airframe changed out of all recognition.

The design was originally intended to be the F9F-8, but the Navy then gave this designation to a straightforward upgrade of the original F9F-6/7 and the supersonic project became the F9F-9. Just as the Panther had become the Cougar, so the new aircraft received its own 'big cat' name from Grumman – Tiger.

The Tiger was to be powered by the Wright J65-W-6, a licence-made version of the British Armstrong Siddeley Sapphire which provided 7700lb of thrust. The Sapphire had been designed without an afterburner in mind but Wright believed that big gains could easily be made by simply adding one. The aircraft would have thin wings and a narrow wingspan of just 31ft 7½in (the Cougar's was 34ft 6in, while the Panther's had been 38ft) but was about 5ft longer than the Cougar at 45ft 10½in.

The fuselage was narrowed at the point where the wings were attached to reduce transonic drag – making the Tiger the first aircraft to be designed from the beginning with the 'wasp waist' or 'Coke bottle' shape.

It had a conventional tail with the tailplanes set low on the fuselage rather than high up like the Panther/Cougar, a sharp-edged fin, and a cockpit which blended smoothly into the fuselage, rather than sitting above it as had previously been the case. The engine's intakes were no longer built into the wingroots but now emerged from the fuselage aft of the cockpit. Armament was four 20mm cannon – two installed below each intake. Underwing pylons would allow the Tiger to carry up to four Sidewinders, multiple rocket pods, a pair of external fuel tanks or a combination of the three.

The first F9F-9 prototype took its first flight on July 30, 1954, without an afterburner but still managed to fly close to the speed of sound. The second prototype, with an afterburner now fitted by Wright, just barely exceeded ▶

GRUMMAN F11F-1 TIGER

Grumman F11F-1 Tiger, NL-107, VA-156, USS *Shangri-la* (CVA-38), 1958. The first variant of the Tiger had a shorter nose with a fixed refuelling probe installed. VA-156 operated the Tiger from 1957 to 1961, being redesignated as VF-111.

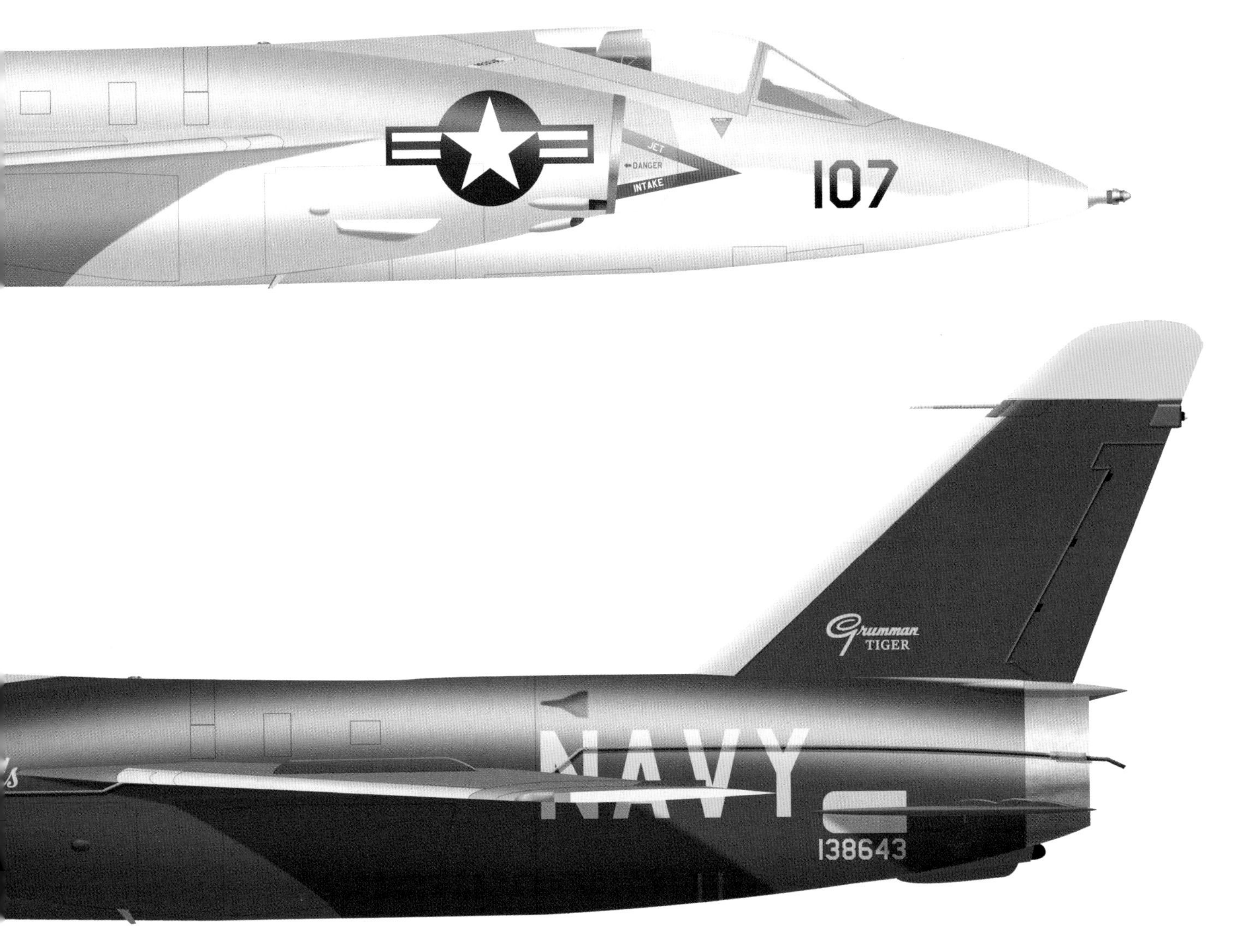

Grumman F11F-1 Tiger

GRUMMAN F11F-1 TIGER

Grumman F11F-1 Tiger, AF-202, VF-33, USS *Intrepid* (CVA-11), 1960. The longer nose version featured a retractable in-flight refuelling probe. VF-33 operated Tigers from 1958 to 1961, the unit being renamed 'Astronauts' during this period.

GRUMMAN F11F-1 TIGER

Grumman F11F-1 Tiger, NJ-115, VF-121, USS *Lexington* (CVA-16), 1958. Tigers were used mainly in the day-fighter role and could carry four AIM-9 Sidewinder AAMs, greatly improving the aircraft's firepower.

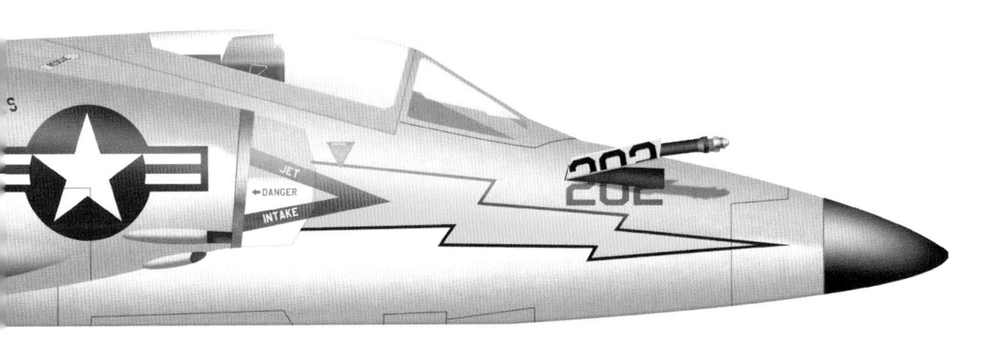

Mach 1 – becoming the second US Navy aircraft to do after the Skyray. Where huge performance gains had been made on other American engines this way, the J65's output only rose to 11,200lb – and the slender aircraft's meagre internal fuel supply was very rapidly exhausted by using the afterburner. Grumman managed to squeeze in additional fuel capacity wherever it could but the aircraft continued to be woefully 'short legged' when it came to range.

Even so, the Navy ordered an initial batch of 42 production aircraft. Early flight testing resulted in a handful of aerodynamic changes. The short and pointed original nose was lengthened somewhat and the rear section of the fuselage underwent a redesign. The Navy was not particularly thrilled by the Tiger but saw sufficient promise to continue supporting the programme.

The new designation F11F-1 was applied in April 1955 and Grumman set about trying to find another engine that would allow the Tiger to realise the full performance potential of its airframe. The ideal candidate turned out to be the new General Electric J79 and work began on redesigning the F11F-1 airframe to accommodate it.

In the meantime, carrier trials with the J65-engined Tiger commenced in April 1956 on board USS *Forrestal*. It was during this phase of testing when, on September 21, 1956, Grumman test pilot Tom Attridge notoriously managed to shoot down his own Tiger. He entered a shallow dive at 20,000ft, accelerating using his afterburner, then fired a four second burst of his four 20mm cannon at 13,000ft. He then fired again to use up his remaining ammunition and continued to dive. At 7000ft his cockpit canopy was hit and cracked but not broken by an external object.

Looking out of his cockpit, he saw a gash in the lip of his right engine intake and found that maximum engine power was reduced to 78%. Two miles from base at 1200ft, with his wheels and flaps down, Attridge realised that he was sinking too quickly and his engine was losing power. He bellied in through a stand of trees a mile from base and lost his right wing in the process, coming to rest before managing to extricate himself from the wrecked aircraft.

Examination of Attridge's stricken Tiger revealed that his fast dive had put him directly into the path of the projectiles he had just fired. One hit his canopy, one the nose cone and another the right engine intake. This last round was found, looking very battered, in the first stage of his engine's compressor.

Deliveries of the 42 short-nose production model Tigers finally commenced in February 1957 with aircraft going to VX-3, a test and evaluation unit. The first fleet squadron to receive them was VA-156 the following ▶

Grumman F11F-1 Tiger

GRUMMAN F11F-1 TIGER

Grumman F11F-1 Tiger, AD-205, VF-21, 1959. The longer nose version was meant to have a radar, although it was never fitted. In 1959, VF-21 was redesignated VA-43 and operated as a fleet replacement squadron for this type.

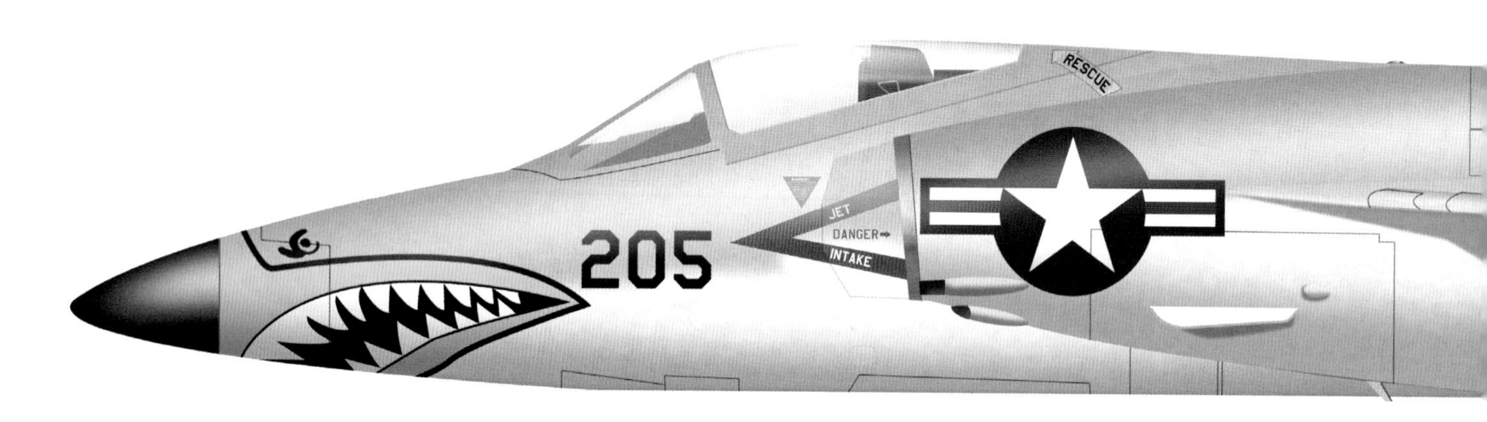

month, redesignated VF-111 shortly thereafter. The Blue Angels display team switched to the F11F-1 at around the same time, though since there were no Tiger trainers a two-seat Cougar F9F-8T as kept on as Number 7.

The Navy now placed a second and final order for 157 further Tigers. These were to have longer noses to house an AN/APS-50 radar, a new J65-W-18 engine and a retractable refuelling probe on the starboard side of the nose – giving the F11F-1 the extra range it so badly needed.

When deliveries of the long-nose Tiger commenced, it was soon apparent that the J65-W-18 was a retrograde step even from the J65-W-6. It provided only 7450lb of dry thrust and just 10,500lb with afterburner, removing the aircraft's ability to go supersonic.

A total of 201 Tigers were built and production was discontinued in December 1958 as more advanced types such as the F8U Crusader became available. The last Tigers were phased out of front line service in April 1961. The Blue Angels, however, kept flying their Tigers until 1968 and the aircraft continued to equip training units until 1967.

Tigers still in service were redesignated F-11A in 1962. Grumman's efforts to re-engine the Tiger with the General Electric J79 in 1955 paid off in the form of the Super Tiger. Fitted with this new powerplant and enlarged intakes, the Tiger airframe, redesignated F11F-1F, managed to reach Mach 2.04 in 1957! It was the first naval aircraft to reach this milestone yet it was never ordered into production due to its lack of carrying capacity.●

GRUMMAN F11F-1 TIGER (F-11)

Grumman F11F-1 Tiger (F-11), 3L-635, VT-26, NAS Chase Field, Texas, 1967. Tigers were operated by training squadrons for supersonic flight training, before pilots converted to their operational fighter type up to the late 1960s. The Tiger was redesignated F-11 after the implementation of the 1962 United States Tri-Service aircraft designation system.

VOUGHT F-8 CRUSADER

1957-1987

The F-8 Crusader was the first Navy jet fighter with real longevity and the last designed with guns as its primary weapon. It was a difficult aircraft to fly but during the Vietnam war it proved to be the best dogfighter in the American arsenal – and a great bomb truck.

VOUGHT F8U-1 (F-8A) CRUSADER

Vought F8U-1 (F-8A) Crusader, NL-408, VF-154, USS *Hancock* (CVA-19), 1958. After the first examples were produced without it, Crusaders were equipped with a retractable in-flight refuelling probe, located on the left side frontal fuselage; in this image it can be seen deployed. VF-154 operated the Crusader from 1957 to 1965.

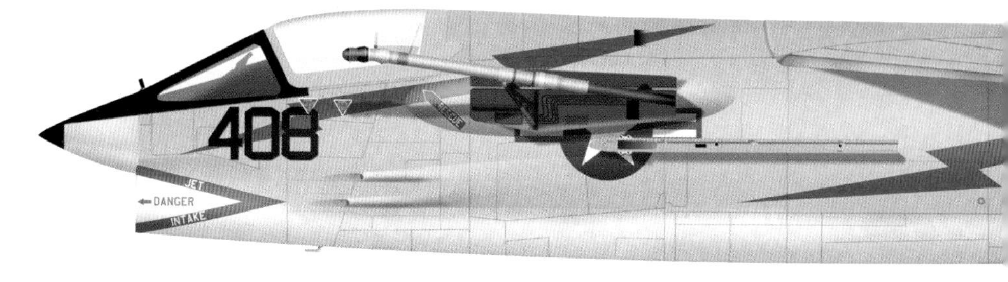

VOUGHT F8U-1 (F-8A) CRUSADER

Vought F8U-1 (F-8A) Crusader, NP-102, VF-211, NAS Miramar, California, 1957. Early Crusaders had a retractable rocket pack with 32 unguided Folding-Fin Aerial Rockets (Mighty Mouse FFARs) – here seen extended. The system presented several problems and was discarded after the C variant. VF-211 operated several variants of the Crusader; it achieved an impressive record of air-to-air kills during the Vietnam War.

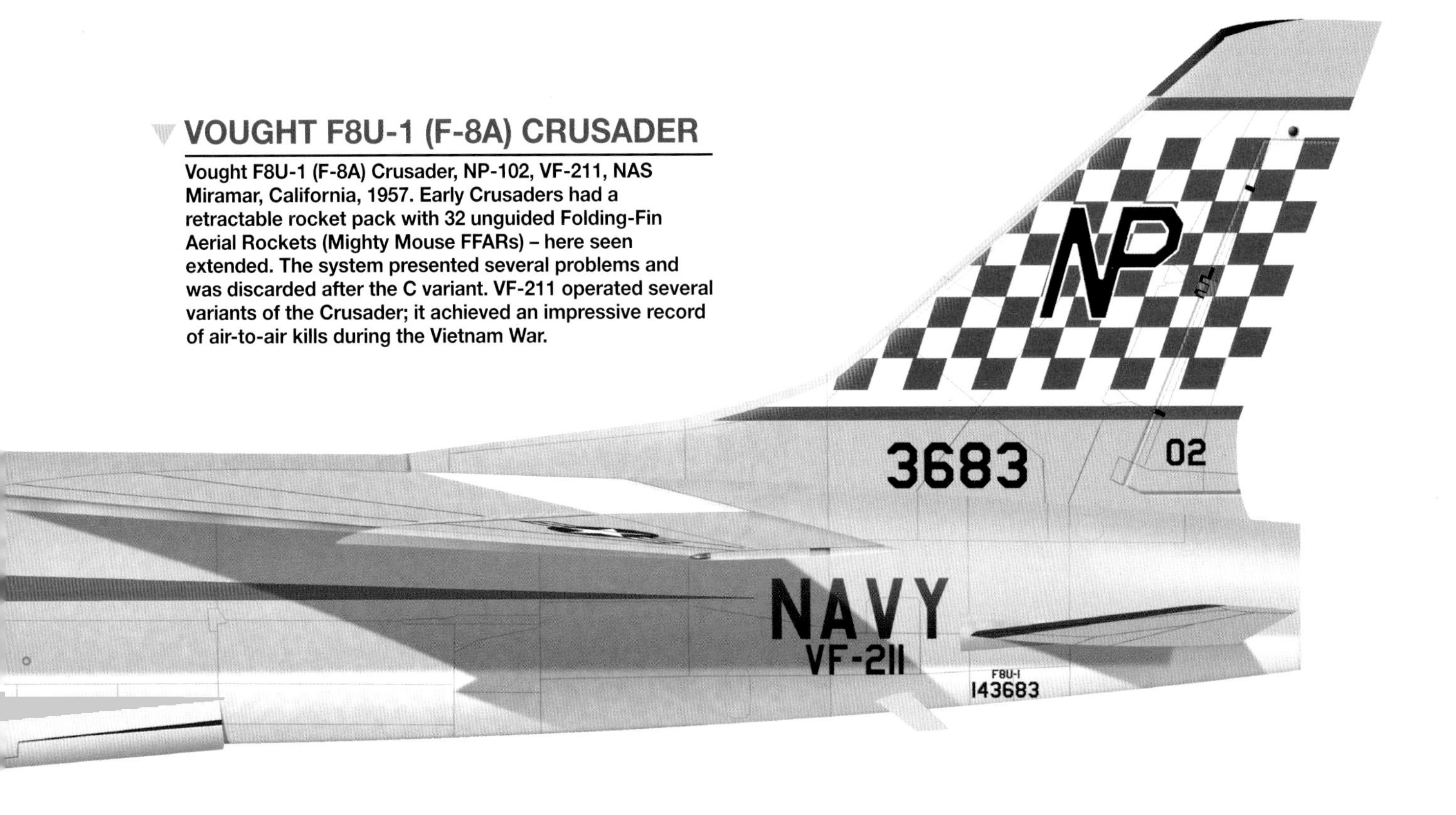

The US Navy remained on the treadmill of commissioning new aircraft with a short service life expectancy well into the 1950s and in September 1952 issued a requirement for a reliable and manoeuvrable new carrier fighter that was also capable of Mach 1.2.

At this time, the Navy was awaiting the redesigned F7U-3 Cutlass from Vought while just beginning to receive deliveries of the still-subsonic Grumman Cougar. The F3H-1 Demon and F4D-1 Skyray were both mired in difficulties arising from the J40 engine and it looked as though one or both might be cancelled.

The Navy needed a next-generation supersonic air superiority fighter and eight manufacturers submitted designs. Among these were the Grumman F11F Tiger, the upgraded McDonnell F3H Demon and a variant of the F-100 Super Sabre which North American called the 'Super Fury'. However, the most promising was Vought's all-new V-383. This rolled all the lessons learned from the Cutlass redesign into a single new aircraft and it appeared to be exactly what the Navy wanted.

Vought had specified the use of both magnesium alloy and titanium, in addition to the usual aluminium alloy, to provide additional strength in crucial structural ▶

Vought F-8 Crusader

areas without increasing weight. The fuselage also featured 'area ruling' – vital for a smooth transition into supersonic flight.

The front of the aircraft was dominated by a large 'chin' air intake for the single Pratt & Whitney J57-P-11 turbojet in the rear fuselage, the latter providing an impressive 9700lb of dry thrust or 14,800lb with afterburner. Further back from the nose were a pair of 20mm Colt Mark 12 cannon on either side of the fuselage. An AN/APG-30 radar gunsight was installed in the nose to help with aiming but there was no full radar system, which meant that the earliest Crusaders would only be able to operate in clear skies.

Behind the guns, attached directly to the fuselage, there was a Sidewinder launch rail. No hardpoints were provided for underwing stores but a retractable rocket pack was installed in the aircraft's underside which could hold 32 2.75in 'Mighty Mouse' unguided folding-fin rockets. However, this feature was seldom used in service. Also on the underside was a large hydraulically actuated air brake.

With a high-wing configuration, all of the landing gear wheels had to retract into the fuselage. The undercarriage legs did not have to be very long in this arrangement and as a result the fighter sat very low to the ground when at rest (in service the Crusader earned the nickname 'Gator', because it crawled around the flight deck on its belly with a big wide open 'mouth' at the front).

The rear part of the fuselage could be detached, providing full access to the engine for servicing or removal, and an arrestor hook was fitted under the tail section which could retract into the fuselage.

The pilot sat high up in the cockpit with excellent visibility upwards, to the sides and down. An ejection seat developed in-house by Vought was provided and there was a pop-out intake on the right side of the fuselage for a Ram Air Turbine or 'RAT'. When activated, slipstream air was sucked in to spin a small turbine which generated emergency back-up power for both the electrical and hydraulic systems.

The most innovative feature in Vought's design, however, was its wings. These were described as 'variable incidence', which in practice meant that the wings sat on top of the fuselage as a single piece, hinged at the rear. During take-off and landing, the front of the wings would be jacked up seven degrees using hydraulics (or a pneumatic back-up if the hydraulics failed).

This meant that while the wings assumed a high angle of attack, the rest of the aircraft remained level, providing the pilot with a much improved forward field of vision. In addition, a linkage which simultaneously lowered the ailerons and leading edge flaps by 25 degrees ▶

VOUGHT F-8C CRUSADER

Vought F-8C Crusader, NP-447, VF-24, USS *Bon Homme Richard* (CVA-31), 1967. This VF-24 aircraft displays MiG kill marks painted on the ventral fins; a testament to the combat record of the squadron during the Vietnam War.

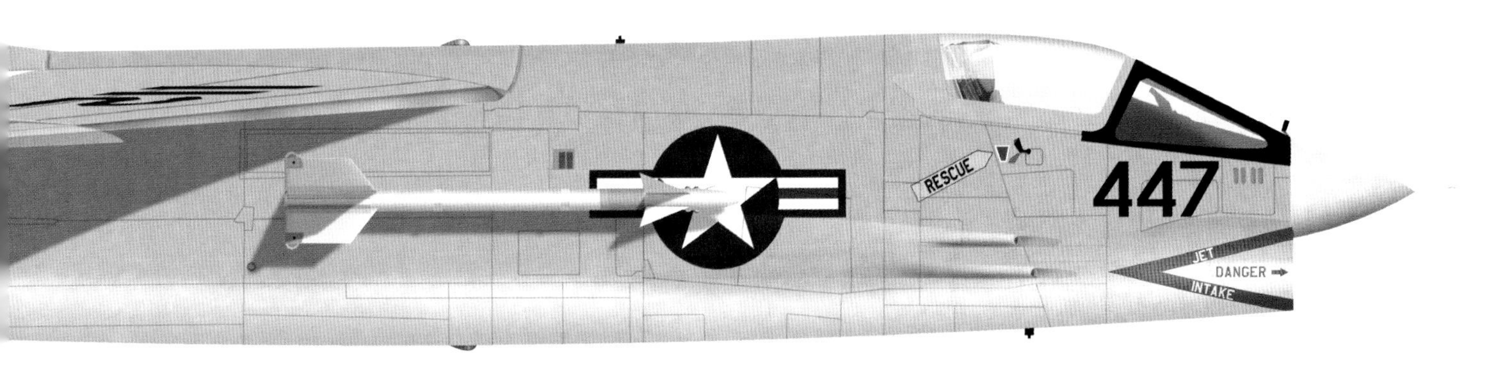

VOUGHT F-8C CRUSADER

Vought F-8C Crusader, AG-209, VF-84, USS *Independence* (CVA-62), 1963. Crusaders had a variable-incidence wing, which improved take-off and landing performance. VF-84 operated the type from 1960 to 1964 before transitioning to the F-4 Phantom II.

Vought F-8 Crusader

VOUGHT F-8D CRUSADER

Vought F-8D Crusader, AJ-106, VF-111, USS *Shangri-La* (CVA-38), 1963. The D variant of the Crusader was an all-weather fighter, featuring improvements to the avionics systems. VF-111 was the first squadron to receive this improved variant. With the Y-shaped pylon (introduced in the C variant) Crusaders could carry two missiles on each of the fuselage hard points; this aircraft carries an AIM-9D IR-guided AAM and an AIM-9C Semi-Active Radar Homing (SARH) AAM.

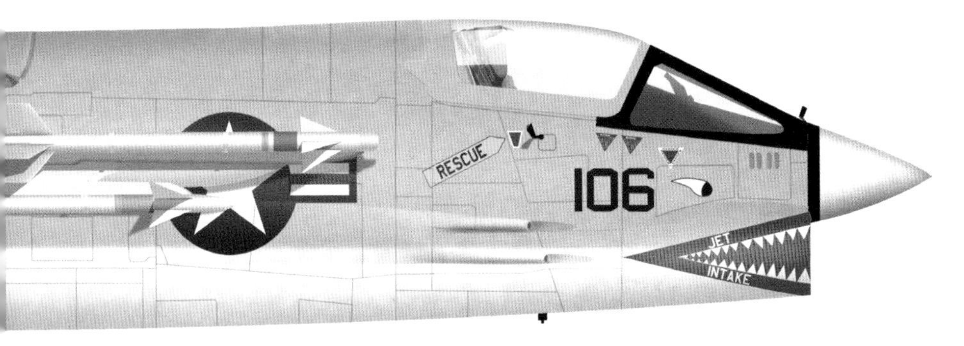

VOUGHT F-8H CRUSADER

Vought F-8H Crusader, AF-201, VF-202, US Naval Reserve, NAS Dallas, Texas, 1975. Several early variants of the Crusader were rebuilt and upgraded, resulting in new designations. F-8Ds were converted into F-8Hs, with the AN/APQ-83 radar and with reinforced landing gear and airframe. Several Crusaders were transferred to US Navy Reserve units, such as VF-202, after their operational career in US Navy front-line squadrons.

decreased take-off and approach speed by providing more lift. The wings had a 5% anhedral and the tips folded up vertically for storage.

The company received a development contract in May 1953 and two prototypes were ordered the following month, on June 29, under the designation XF8U-1. The first of these was rolled out in February 1955 and Vought chief experimental test pilot John W Konrad managed to take it above Mach 1 during its first flight on March 25. The second prototype made its flight debut on June 12 and the name 'Crusader' was allocated by the company.

Testing proceeded swiftly and failed to reveal any serious problems. As a result, the first production F8U-1 Crusader rolled out on September 20, 1955. This initially had the slightly improved J57-P-12 turbojet but after the 31st aircraft production was switched to the more powerful J57-P-4A, with 10,900lb of dry thrust – an increase of 800lb – and 16,000lb with afterburner, up 1200lb on the -11 and -12.

Carrier qualification trials were carried out on USS *Forrestal* from January to April 1956 and deliveries to squadrons commenced on December 28, 1956, with a total of 318 F8U-1s being built.

The 32nd of these was modified to become a prototype for the F8U-1P reconnaissance version. All armament and associated systems were stripped out and replaced with a series of five camera stations under the forward fuselage. These required two small square windows on either side of the aircraft plus another three on the underside and a fourth in a forward-facing pod directly underneath the cockpit. The aircraft's tailfin was also decreased in height to improve aerodynamics. The prototype first flew in its converted form on December 17, 1957, and a total of 144 F8U-1Ps were built.

By this time there were already numerous US Navy and Marine squadrons equipped with the Crusader. The type's first carrier deployments took place during early 1958 on the USS *Saratoga* in the Atlantic and the USS *Hancock* in the Pacific. That July, during the Lebanon Crisis, Crusaders from the former provided top cover during the US intervention without encountering any opposition.

Pilots flying the Crusader were pleasantly surprised to discover that it actually exceeded the manufacturer's promised performance– managing Mach 1.7 against Vought's specification of Mach 1.4. However, it could be difficult to recover from a spin and also difficult to land, since despite the variable incidence wing and other aids it still had a high approach speed. Pilots also had to remember to retract the air brake – though this sometimes hung open anyway due to ▶

Vought F-8 Crusader

leaky hydraulics. As a result of all these factors, the Crusader had a very high accident rate – with 87% of all those built being involved in a mishap at some point.

Following on from the F8U-1 was the improved F8U-1E, which offered limited all-weather capability by swapping the original's AN/APG-30 radar gunsight for an AN/APS-67 radar system. A larger nosecone was required but otherwise the F8U-1E was very similar to the F8U-1. The first example flew in early September 1958 and a total of 130 were built.

'CRUSADER II'

In parallel to the F8U-1E, Vought was working on what it referred to as the 'Crusader II': the F8U-2. This included a new engine, the J57-P-16, which provided slightly less dry thrust at 10,700lb but increased thrust with afterburner, up to 16,900lb, making the aircraft capable of approaching Mach 2.

The original Sidewinder racks were replaced with new Y-shaped ones – allowing the aircraft to carry two missiles on each side rather than just one – but in service these aircraft still usually only carried one on either side.

Other new features included a Martin-Baker Mark 5 ejection seat (which did not offer zero-zero capability), replacing the Vought design, a pair of intakes on top of the tailcone for afterburner cooling and two ventral strakes on the underside of the tail for improved stability. Wingspan was decreased by 6in as well.

A prototype for the F8U-2 was created by stripping out a standard F8U-1 and installing the new engine. The aircraft first flew in this configuration in December 1957 and a more fully modified second prototype flew the following month. The first production example made its debut in August 1958 and by September 1960 a total of 187 had been built.

Next came a Crusader night-fighter, with another new engine – the J57-P-20, providing a staggering 18,000lb thrust with the afterburner on. The rocket belly pack, long since sealed up on previous Crusader marks, was now deleted to provide additional internal fuel capacity and improved radar and fire control systems were installed, plus an autopilot. An infrared search and track sensor, which appeared externally as a nodule just above the aircraft's nosecone, was also fitted. First flight of the F8U-2N was in February 1960, production deliveries commenced four months later and by the beginning of 1962 a total of 152 had been built.

The final new-build US Navy Crusader was the F8U-2NE. This had the new AN/APQ-94 search and fire control radar system fitted, providing even greater night and all-weather capability. It also had, for the first time on a Crusader, the option to fit a pair of underwing pylons for external stores. These could carry up to 5000lb of ordnance including rockets, ▶

VOUGHT F-8K CRUSADER

Vought F-8K Crusader, JH-2, VCF-10, NAS Guantanamo Bay, Cuba, 1971. Although US Navy Crusaders were tasked mainly with an air-to-air role, the F-8 could be equipped with air-to-ground weapons such as these Zuni rockets carried in twin-pods on the fuselage hard points and MK-82 Snakeye bombs on Multiple-Ejector Racks on the wing pylons. The F-8Ks were upgraded F-8Cs featuring improvements in weapons capability, engine and radar. VCF-10 was a composite squadron flying different types of aircraft.

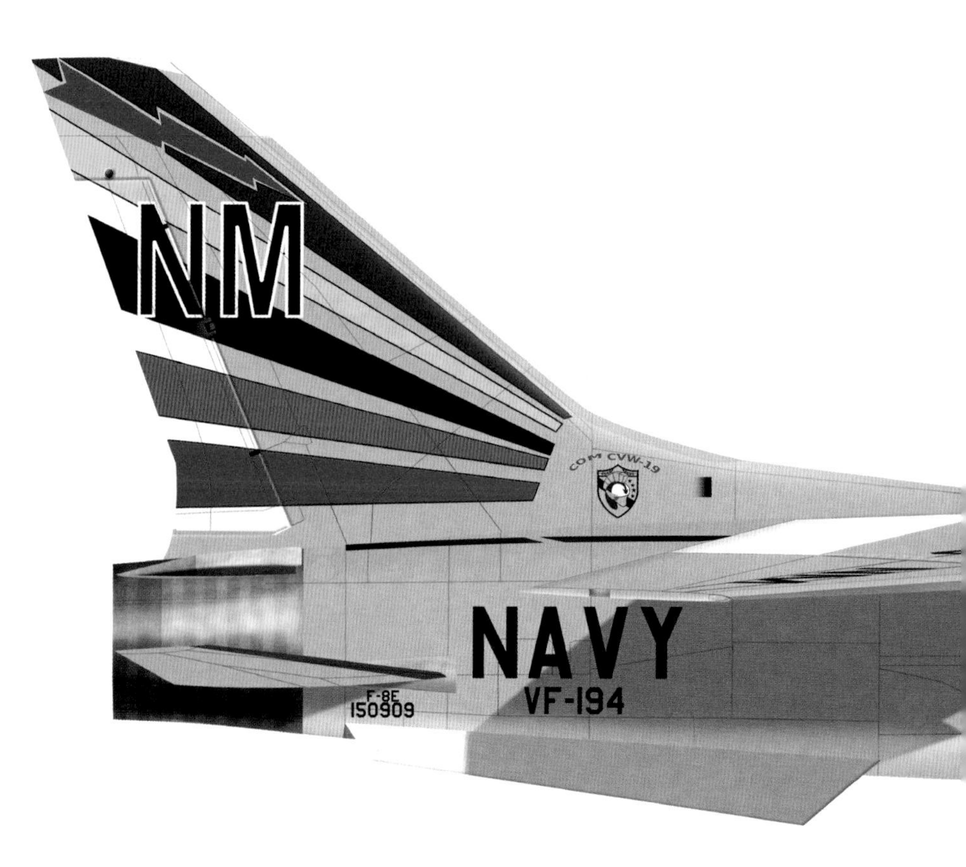

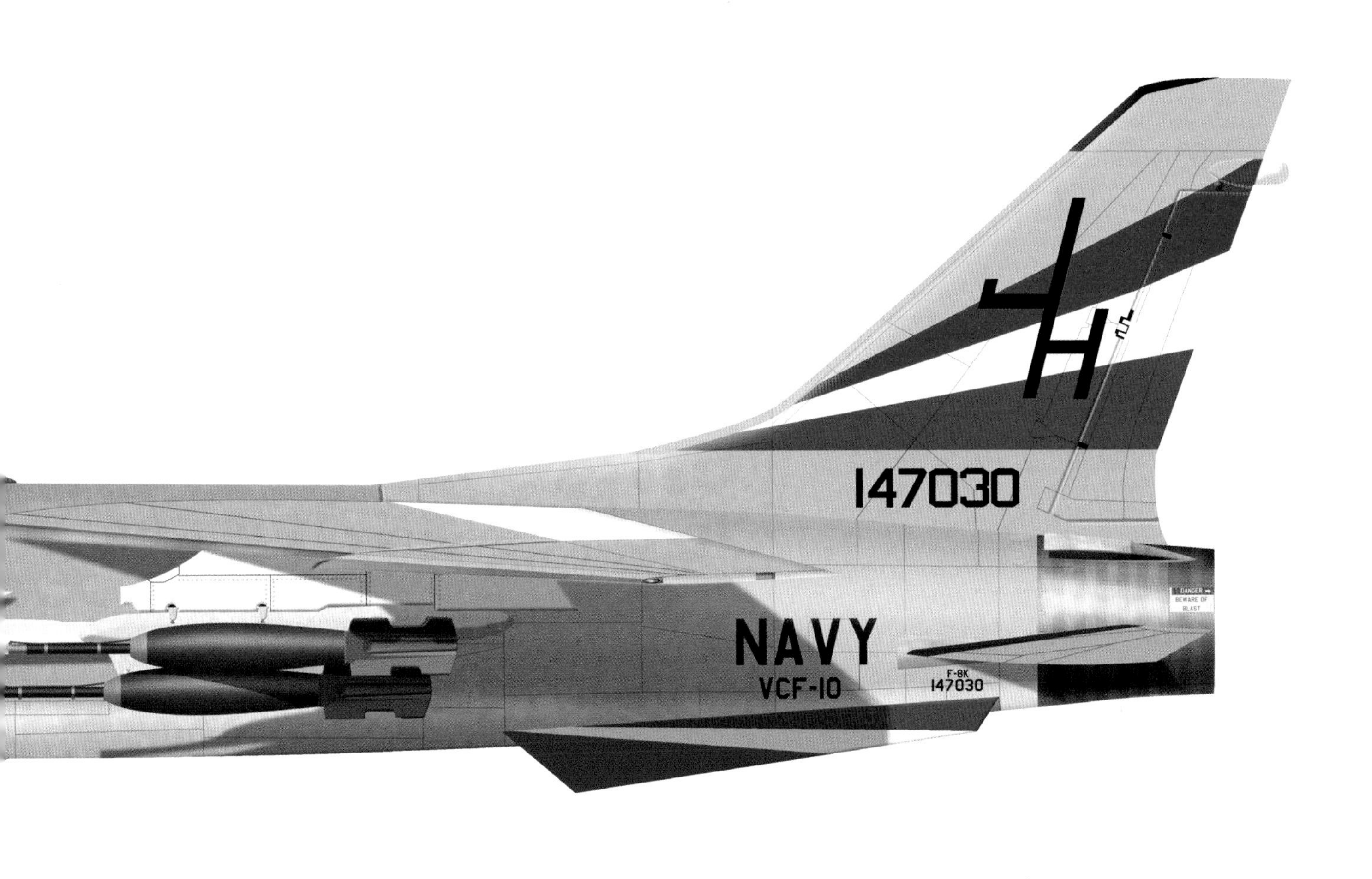

VOUGHT F-8E CRUSADER

Vought F-8E Crusader, NM-400, VF-194, USS *Ticonderoga* (CV-14), 1966. The official aircraft of the Commanding Officer of US Navy Carrier Groups, known as the CAG Bird, is normally the most visually striking machine aboard; these aircraft have a number usually ending in double zero. VF-194 was equipped with Crusaders in 1959 when it was designated as VF-91; it continued to operate F-8s until 1976, being among the last US Navy fighter squadrons to do so.

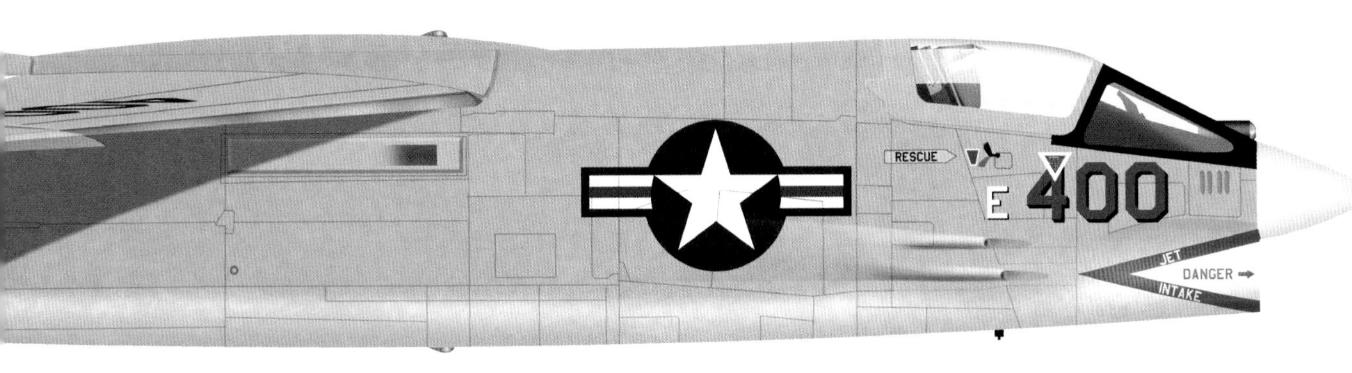

Vought F-8 Crusader

VOUGHT DF-8A CRUSADER

Vought DF-8A Crusader, GF-28, VC-8, NAS Roosevelt Roads, Puerto Rico, 1970. This colourful Crusader is a DF-8A variant, a specially-modified F-8A for the drone-control role.

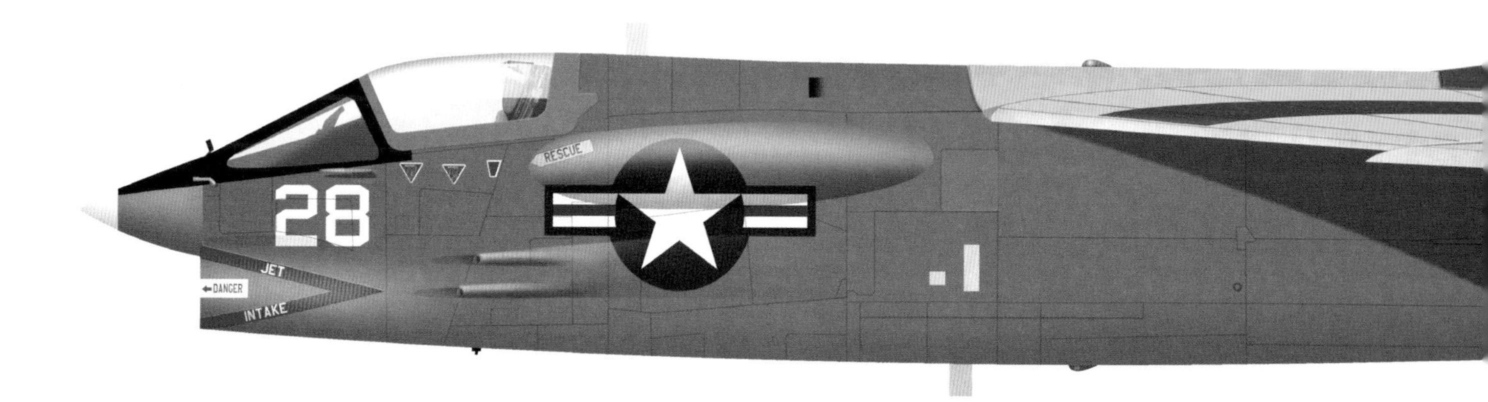

VOUGHT F-8J CRUSADER

Vought F-8J Crusader, NM-112, VF-191, USS *Oriskany* (CV-34), 1972. F-8Js were rebuilt F-8Es, featuring improvements to the engine and radar and wet pylons allowing for the carriage of external tanks (although tested, these were not carried operationally). VF-191 was equipped with several variants of the Crusader from 1960 to 1976, being one of the last US Navy fighter squadron to fly the F-8; it made several deployments to the Vietnam theatre of war from 1964 to 1974.

bombs and air-to-surface missiles – specifically the AGM-12 Bullpup, for which guidance systems were installed in a centre-wing hump. A total of 286 F8U-2NEs were constructed, the largest of all the Crusader production runs after the original F8U-1.

In order to assist with the difficulties pilots were having in transitioning to the Crusader from less demanding types, Vought took the 77th production model F8U-1 and turned it into the F8U-1T 'Twosader' trainer with an additional cockpit installed. This required an entirely new forward fuselage that was 2ft longer than the original, with two of the cannon and the rocket pack removed. It also featured a drag chute tucked into an installation at the base of the fin which effectively halved the distance usually required to land. The sole Twosader first flew in early 1962 but no production orders were forthcoming.

Vought tried to sell the two-seater to Britain as the basis for a new naval fighter but the British government decided to purchase a variant of the Phantom fitted with the UK-made Rolls-Royce Spey engine instead.

In common with all other US Navy fighters, the various F8U models were redesignated in September 1962. The F8U-1 became the F-8A, the F8U-1P became the RF-8A, the F8U-1E was the F-8B, the F8U-2 became the F-8C, the F8U-2N became the F-8D and the F8U-2NE became the F-8E.

From 1963, the Vought Crusader became the Ling-Temco-Vought Crusader after Ling-Temco acquired Vought in a hostile takeover.

REBUILDS

All surviving Crusaders were retrofitted with an auto-throttle from 1964 to smooth out the aircraft's throttle response at low speeds and make it easier to land and starting in 1965 an extensive programme of renovation and refurbishment was carried out on the remaining F-8s. This included stronger new wings with the fitment of pylons as a universal option plus new landing gear based on that of the Vought A-7A Corsair II with a longer nosewheel strut.

The programme did not include any standard F-8As but the other types were redesignated as follows: the RF-8A became the RF-8G (73 aircraft), the F-8D became the F-8H (89 aircraft), the F-8E became the F-8J (136 aircraft), the F-8C became the F-8K (87 aircraft) and the F-8B became the F-8L (61 aircraft).

In 1968 the newer Crusaders were fitted with Martin-Baker Mark 7 ejection seats which finally gave the aircraft's pilots zero-zero ejection as an emergency option. A number of the old F-8s were converted into drone controllers as DF-8As and drones as QF-8As and DQF-8As in the late 1960s.

Vought F-8 Crusader

VOUGHT F8U-1P (RF-8A) CRUSADER

Vought F8U-1P (RF-8A) Crusader, GA-910, VFP-62, NAS Cecil Field, Florida, 1962. RF-8A was the new designation for the F8U-1P in 1962; based on the F-8A, the aircraft featured changes to the front fuselage, allowing the carriage of multiple cameras. VFP-62 distinguished itself during the Cuban Missile Crisis. Operating from Florida, it carried out low-level reconnaissance missions over Cuba, gathering photographic evidence of the Soviet military build-up on the island (Operation Blue Moon). This aircraft carries the Navy Unit Commendation ribbon on its nose.

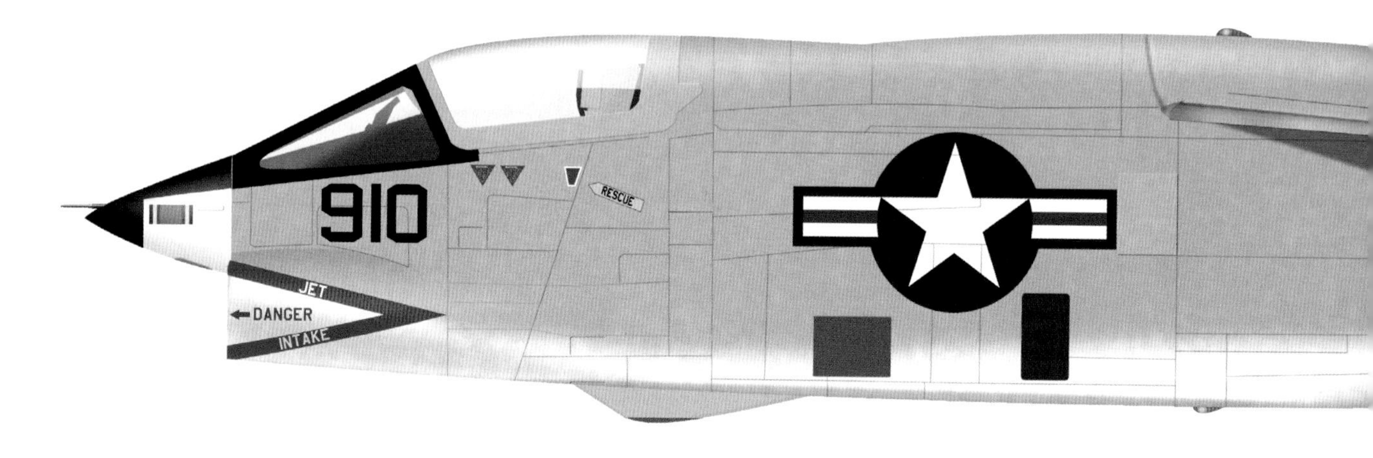

The Crusader had reached the peak of its development just as the Vietnam War was beginning and the fighter would serve with distinction throughout the conflict – from overflights in 1964 to strike missions during Operation Rolling Thunder and, most famously, dogfights in 1966-68. The F-8, the 'last of the gunfighters', demonstrated exactly what a high-performance fighter armed with four 20mm cannon and Sidewinders could do by shooting down 18 MiGs for the loss of four Crusaders in air-to-air combat.

Flak and surface-to-air missiles would take a much heavier toll on US Navy Crusaders, however, with 42 F-8s and 20 RF-8s being lost in Vietnam due to the former and 10 F-8s to the latter. A further 58 F-8s and nine RF-8s were wrecked in accidents or rendered inoperable by mechanical failure.

The replacement of F-8s by F-4s commenced in 1972 and the last of the fighter variants was retired in May 1976. The RF-8Gs continued to serve into the early 80s, with ongoing upgrades, until those remaining were finally moved to the Navy Reserve in May 1982. The last RF-8Gs in the Reserve were retired in March 1987, being replaced by F-14s with the TARPS recon pod.●

VOUGHT RF-8G CRUSADER

Vought RF-8G Crusader, AF-702, VFP-206, US Navy Reserve, NAS Washington, Maryland, 1987. Many Crusaders ended their operational career with the US Navy Reserve, well into the 1980s; these aircraft featured grey camouflage paint and low-visibility markings. RF-8G variants were upgraded RF-8As, featuring ventral fins and new electronic equipment. VFP-206 was the last US Navy unit to operate the Crusader.

MCDONNELL DOUGLAS F-4 PHANTOM II

The iconic F-4 Phantom II was begun as an attempt to update and improve the underperforming F3H Demon – but it would go on to become one of the most successful jet fighters in history.

1960-2004

MCDONNELL F4H-1F (F-4A) PHANTOM II

McDonnell F4H-1F (F-4A) Phantom II, NJ-102, VF-121, NAS Miramar, California, 1961. VF-121 became the first US Navy Squadron to receive the new fighter. The F-4 was designated F4H-1 until the implementation of 1962's United States Tri-Service aircraft designation system.

As the XF3H-1 prototypes commenced carrier trials in August 1953, McDonnell preliminary design manager David S Lewis began to consider a series of concepts for upgrading of the Demon.

These were designated Model 98A, B, C, D and E. Model 98A would see the Demon fitted with a Wright J67 in place of its Westinghouse J40 – with the goal of producing a variant capable of hitting Mach 1.69.

The slightly more radical Model 98B was designed to perform as either a fighter or a photo reconnaissance platform and two further engine options were pencilled in – a pair of Wright J65s or a pair of General Electric J79s. An enlarged wing was provided to account for the additional weight.

Model 98C had the same choice of engines as 98B but was fitted with a completely new delta wing – though it would still have a conventional tail – whereas Model 98D had a new straight wing.

Finally, Model 98E also had a delta wing and conventional tail but the wing was larger and thinner.

In order to make these concepts more attractive, Lewis deliberately made them adaptable. Different nose sections could be fitted depending on the desired role, the fuselage could be produced as either a single- or two-seater and nine hardpoints offered massive flexibility on loadout.

The company believed that these designs had potential and submitted them to the Navy's Bureau of Aeronautics as an unsolicited proposal for a 'Super Demon' on September 19, 1953.

The Navy was most interested in Model 98B and gave the company a contract to produce a full-scale mock-up in early 1954. The variant initially studied was a single-seater with two engines and a delta wing with a 45° sweepback on its leading edge.

The design was subsequently reworked with 11 hardpoints in order to make the fighter a fighter-bomber, since the Navy believed that the F11F Tiger and Vought F-8 Crusader would fulfil the interceptor role.

Two prototypes were ordered following an inspection of the mock-up on October 18, 1954, as ground-attack aircraft under the designation YAH-1. However, this role was then taken by the Douglas A-4 Skyhawk.

In the meantime, the F3H-2N Demon was beginning to enter fleet service as an all-weather missile-armed fleet defence fighter – so the 'Super Demon' project was repurposed to become its successor.

The design was switched to the two-seat variant and redesignated YF4H-1, ▶

MCDONNELL F4H-1F (F-4A)

McDonnell F4H-1F (F-4A) AD-188, VF-101, Detachment A, Los Angeles, California to New York, 1961. On May 24, 1961, aircraft from VF-101 flew from Los Angeles to New York in two hours and 47 seconds, winning the Bendix Trophy. Naval Phantoms used a probe and drogue refuelling system with a retractable probe installed on the right side fuselage.

McDonnell Douglas F-4 Phantom II

MCDONNELL F-4B PHANTOM II

McDonnell F-4B Phantom II, NE-101, VF-21, USS *Midway* (CV-41), 1965. In 1965, aircraft from VF-21 became the first to score kills on North Vietnamese MiGs; on June 17, 1965, two aircraft from the squadron destroyed MiG-17s using AIM-7 Sparrow AAMs.

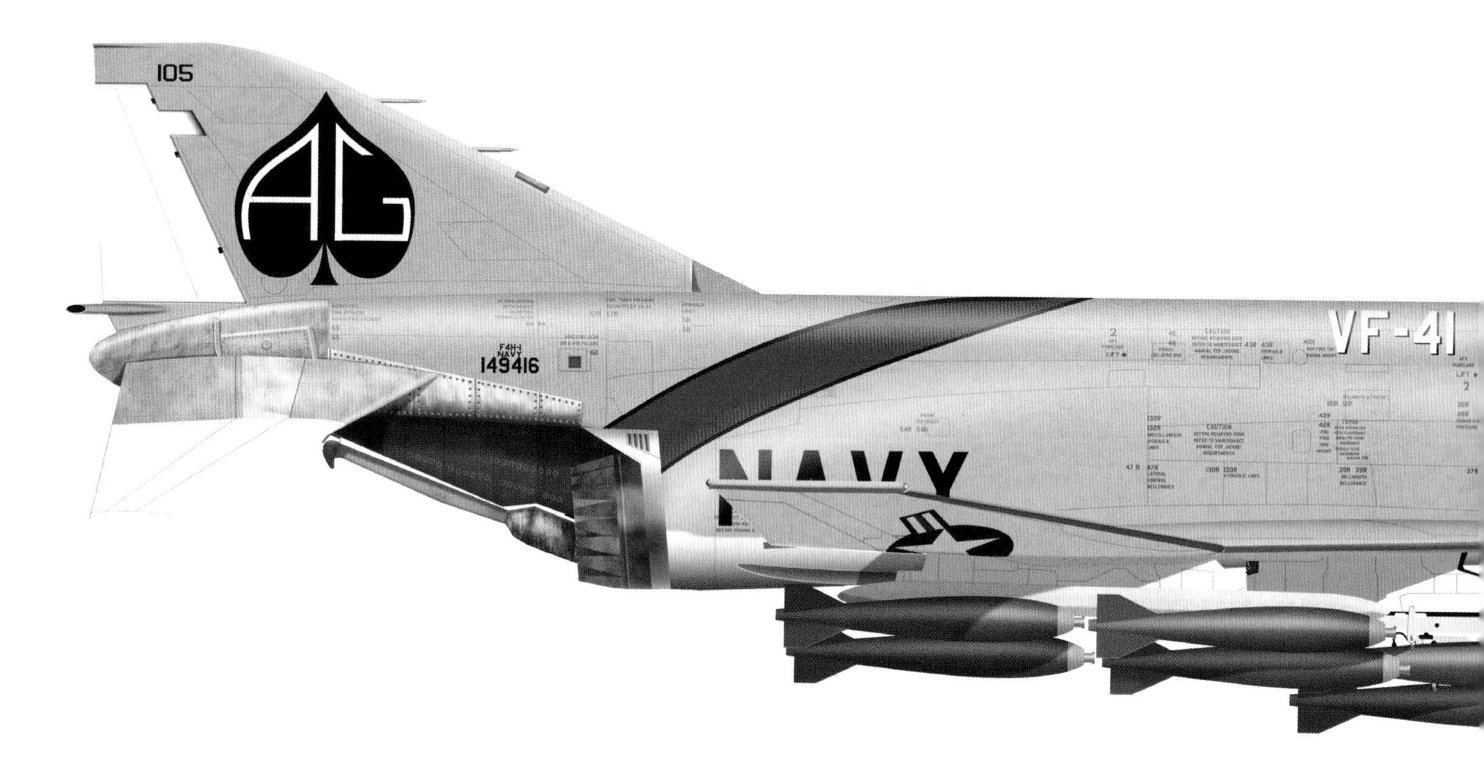

MCDONNELL F-4B PHANTOM II

McDonnell F-4B Phantom II, NF-102, VF-161, USS *Midway* (CV-41), 1973. On January 1973, an aircraft from VF-161 would score the final (197th) MiG kill of the Vietnam War, a MiG-17 destroyed using an AIM-9 Sidewinder AAM.

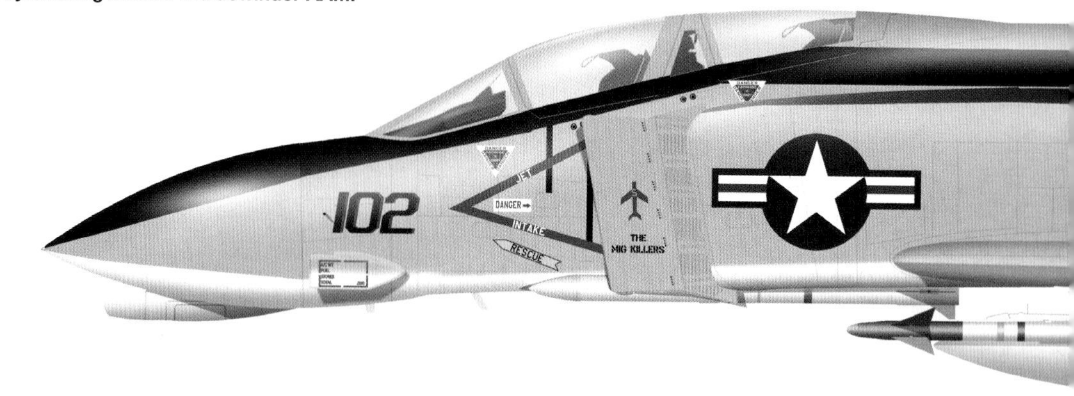

MCDONNELL F4H-1 (F-4B) PHANTOM II

McDonnell F4H-1 (F-4B) Phantom II, AG-105, VF-41, NAS Lemoore, California, 1962. Although designed primarily as a fleet defence aircraft, the Phantom could also carry air-to-ground weaponry; here a VF-41 aircraft carries MK.82 bombs. This squadron would operate Phantoms until 1976 when it transitioned to the new F-14 Tomcat.

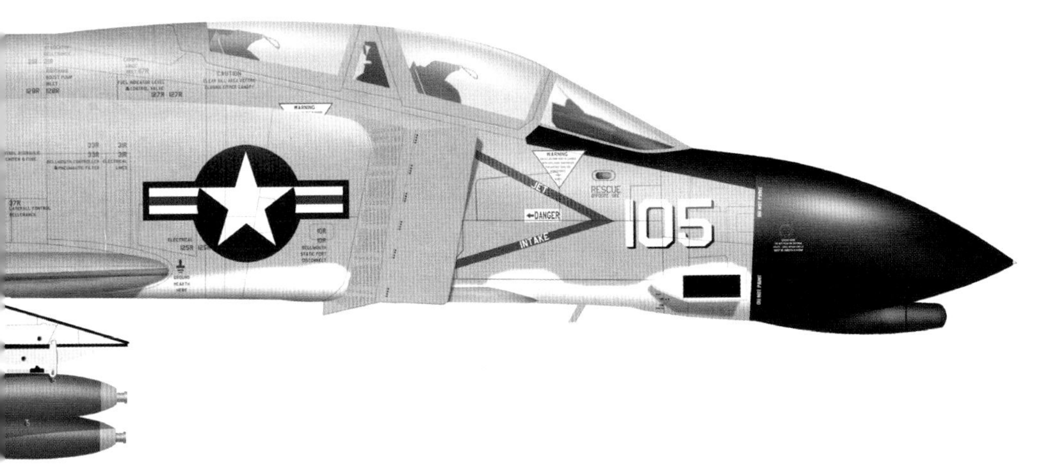

McDonnell Douglas F-4 Phantom II

with General Electric J79-GE-3As being chosen to power it.

The engines were positioned low down in the rear fuselage and their intakes were positioned on the sides of the fuselage next to the back-seater's position. The wing was a thin-section delta with a leading edge sweep of 45° and the outer wings were angled up by 12° to cure a lateral instability problem revealed during wind tunnel testing. A 'dogtooth' was added to the leading edge for better control at high angles of attack and for the same reason the tailplanes were given a 23.25° anhedral.

There would be no internal cannon required for the fleet defence role, so armament was to be four AIM-7 Sparrows in semi-recessed bays under the fuselage with the option to supplement them with Sidewinders on underwing pylons. An AN/APG-50 radar unit was installed in the aircraft's nose for missile guidance. The YF4H-1 would become the first US fighter armed solely with missiles.

Robert C Little took the YF4H-1 Phantom II up for its first flight on May 27, 1958. Early testing revealed that modifications to the intakes were necessary – with the introduction of 12,500 little holes in the inner intake door to extract boundary layer air. A new system was also introduced whereby air from the compressors was blown across the leading edge slats

MCDONNELL F-4B(G) PHANTOM II

McDonnell F-4B(G) Phantom II, NJ-111, VF-121, NAS Miramar, California, 1966. In 1963, the US Navy used 12 modified F-4Bs to conduct experiments on automatic carrier landings; the aircraft were modified with an air-to-ground data link, an approach power compensator and a radar reflector installed on the nose. Later, in 1966, these aircraft were also painted in a new experimental green upper surfaces camouflage. Neither the automatic landing system nor the new camouflage were adopted for service and these aircraft were returned to their regular configurations.

MCDONNELL DOUGLAS F-4J PHANTOM II

McDonnell Douglas F-4J Phantom II, NG-100, VF-96, USS *Constellation* (CVA-64), 1972. In May of 1972, Lt Randy Cunningham and Lt (jg) Willie Driscoll would score their fifth air-to-air kill and achieve ace status; kills three, four and five were made on the same day: May 10, 1972.

and trailing edge flaps to improve their performance.

Seven months after Little's first Phantom flight, in December 1958, the type defeated the rival XF8U-3 Crusader III in comparative flight tests. An initial order for 20 pre-production model F4H-1s was now increased to 45, which would be fitted with the J79-GE-2 or -2A – a less powerful version of the J79-GE-8 planned for the full production model. In the meantime, the design underwent further changes – the cockpit canopy was raised up to improve visibility for the crew, two new ramps were added to the intake ducts – one fixed and one variable – and the nose was increased in diameter to house the AN/APQ-72 radar.

The second YF4H-1 prototype was used to set a world record altitude of 98,557ft as part of Project Top Flight on December 6, 1959, and on February 15, 1960 the fourth F4H-1F made the Phantom's first aircraft carrier launch and recovery aboard USS *Independence*. The pre-production Phantoms had 'F' added to their designation to differentiate them from the full production model F4H-1 which followed.

The F4H-1 had J79-GE-8 engines, with 17,000lb of thrust in full afterburner, and the aircraft's intake ramps were modified again to suit them. It also had improved internal fuel capacity thanks to space-saving refinements within the fuselage. In addition to the four ▶

McDonnell Douglas F-4 Phantom II

Sparrows, the F4H-1 could carry four rail-mounted AIM-9 Sidewinders on the inner wing pylons.

The first operational unit to receive the Phantom was VF-74, who got their first example on July 8, 1961. A new low altitude speed record of 902.769mph was set by Lt H Hardisty and Lt E H DeEsch using an F4H-1F during Operation Sageburner on August 28. Three months earlier, on May 18, Cdr J L Felsman had been killed when his F4H-1F disintegrated following a pitch damper failure during the first attempt at Operation Sageburner. It was the first fatal Phantom II accident.

Twenty-nine F4H-1s were loaned to the USAF for trials in October 1961, the first two being marked 'F-110A', and during the same month VF-74 became the first Phantom squadron to complete its carrier qualifications. VF-114 also became operational with the Phantom.

On November 22, 1961, US Navy pilot Lt Col Robert B Robinson broke the absolute world air speed record in the second YF4H-1 during Operation Skyburner at Edwards AFB – reaching 1606.3mph, the first record to be set above Mach 2. The same aircraft was used to set the world record for sustained altitude less than two weeks later when Cdr G W Ellis managed 66,443.8ft above Edwards on December 5.

Eight time-to-climb records were then set during Operation High Jump between February and April 1962, by various pilots using various production F4H-1s.

MCDONNELL DOUGLAS F-4J PHANTOM II

McDonnell Douglas F-4J Phantom II, XF-9, VX-4, NAS Point Mogu, California, 1969. Air Test and Evaluation Squadron 4 (VX-4) used several variants of the Phantom, from the early 1960s up until 1990. It is part of the Operational Test and Evaluation Force (OPTEVFOR).

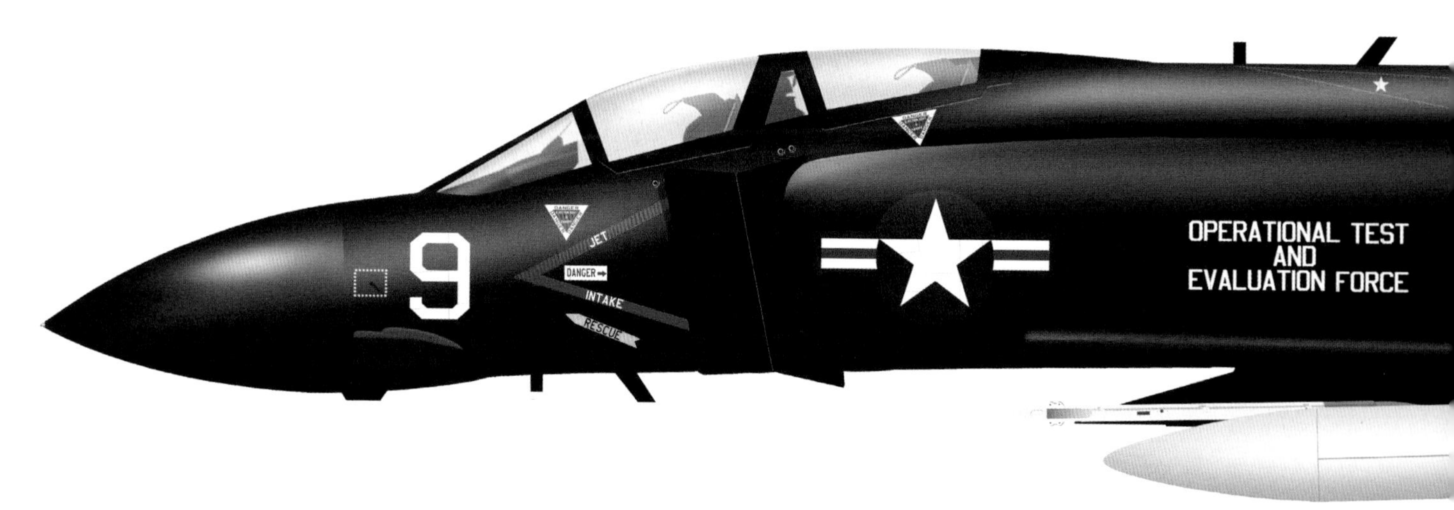

In September 1962, the remaining F4H-1Fs were given the new designation F-4A and the F4H-1s became F-4Bs. The following month VF-102 was deployed with a full complement of Phantoms aboard USS *Enterprise*. VF-74 was deployed aboard USS *Forrestal* between August 1962 and March 1963.

US Navy pilots from VF-142 and VF-143 aboard USS *Constellation* flew escort for A-4 Skyhawks during Operation Pierce Arrow against North Vietnamese gunboats in the Gulf of Tonkin on August 5, 1964: the first F-4B combat missions. And the first 'kill' by a Phantom happened on April 9, 1965, when a VF-96 F-4B flown by Lt (jg) Terrence M Murphy and Ensign Ronald J Fegan, flying from USS *Ranger*, with call-sign 'Showtime 602' shot down a Chinese MiG-17. Shortly afterwards Murphy and Fegan's F-4B disappeared and the pair were never seen again. Parts of the mission record remain classified but it has been speculated that they were either shot down by another MiG or an AIM-7 missile from one of their own wingmen.

An F-4B of VF-21 'Freelancers' flown by Cdr Louis Page and Lt John C Smith shot down the first North Vietnamese MiG of the war on June 17, 1965. A total of 649 F-4Bs were built for the US Navy and Marine Corps and numerous upgrades were made during the war. Defensive capabilities were improved with the AN/ALQ-51 and -100 radar jamming and track breaking systems, chaff and ▶

MCDONNELL DOUGLAS F-4J PHANTOM II

McDonnell Douglas F-4J Phantom II, VX-4, NAS Point Mogu, California, 1976. During this year several US military aircraft were painted in special liveries to celebrate the 200th anniversary of the American revolution, including this Phantom from the Air Test and Evaluation Squadron 4.

McDonnell Douglas F-4 Phantom II

MCDONNELL DOUGLAS F-4J PHANTOM II

McDonnell Douglas F-4J Phantom II, 1, Blue Angels, NAS Pensacola, Florida, 1970. The famous demonstration team flew the Phantom from 1969 to 1974 before adopting the A-4 Skyhawk. The F-4 Phantom was the only aircraft flown by both the Blue Angels and the USAF's Thunderbirds.

MCDONNELL DOUGLAS YF-4J PHANTOM II

McDonnell Douglas YF-4J Phantom II, 1473, NWCS, Naval Weapons Center, NAS Point Mugu, California, 1987. Originally built as an F-4B, this aircraft was converted into an YF-4J prototype and used in several test programmes, including that of the new NACES ejection seat.

MCDONNELL DOUGLAS F-4N PHANTOM II

McDonnell Douglas F-4N Phantom II, NL-200, VF-111, USS *Coral Sea* (C-43), 1975. The F-4Ns were modernised F-4Bs, featuring improvements in structure, aerodynamics and engines. This VF-111 aircraft is the CAG bird and displays a suitable colour scheme.

McDonnell Douglas F-4 Phantom II

MCDONNELL DOUGLAS F-4N PHANTOM II

McDonnell Douglas F-4N Phantom II, 94, PMTC, NAS Point Mugu, 1984. This Phantom from the Point Mugu Test Center carries the AQM-81N Firebolt drone.

flare dispensers and a radar homing and warning system.

A small number of F-4Bs were modified to become DF-4B drone directors and NF-4B testbeds. The EF-4B was an electronic warfare conversion with jamming pods and other underwing countermeasures systems in the early 1970s. However, many ended their service careers as QF-4B supersonic target drones. Finally, 46 became RF-4B reconnaissance aircraft for the Marine Corps. These had their radar noses removed and a new camera nose added that was 4ft 8.875in longer.

THE F-4J, N AND S

Even as the Phantom was getting to grips with the enemy in Vietnam, McDonnell was hard at work on a new Navy variant. The F-4J had new engines – the J79-GE-10 with 17,844lb-ft of thrust each with afterburner – but this was just the start. The nose was enlarged to accommodate the Westinghouse AN/APG-59 pulse doppler radar with integrated AN/AWG-10 Fire Control System (FCS) for look-down, shoot-down capability – which meant it could pick out and track low flying targets.

Data generated by the FCS was fed directly to the pilot via a helmet-mounted sight and another change in the cockpit

MCDONNELL DOUGLAS F-4N PHANTOM II

McDonnell Douglas F-4N Phantom II, AF-201, VF-202, US Naval Reserve, NAS Dallas, Texas, 1978. Although replaced by newer aircraft in both the US Navy and USAF, Phantoms could still hold their own against these new machines, especially when flown by more experienced crews. This US Naval Reserve aircraft displays markings of those achievements in its splitter plate, hailing the results of mock air combats against the USAF's new F-15s.

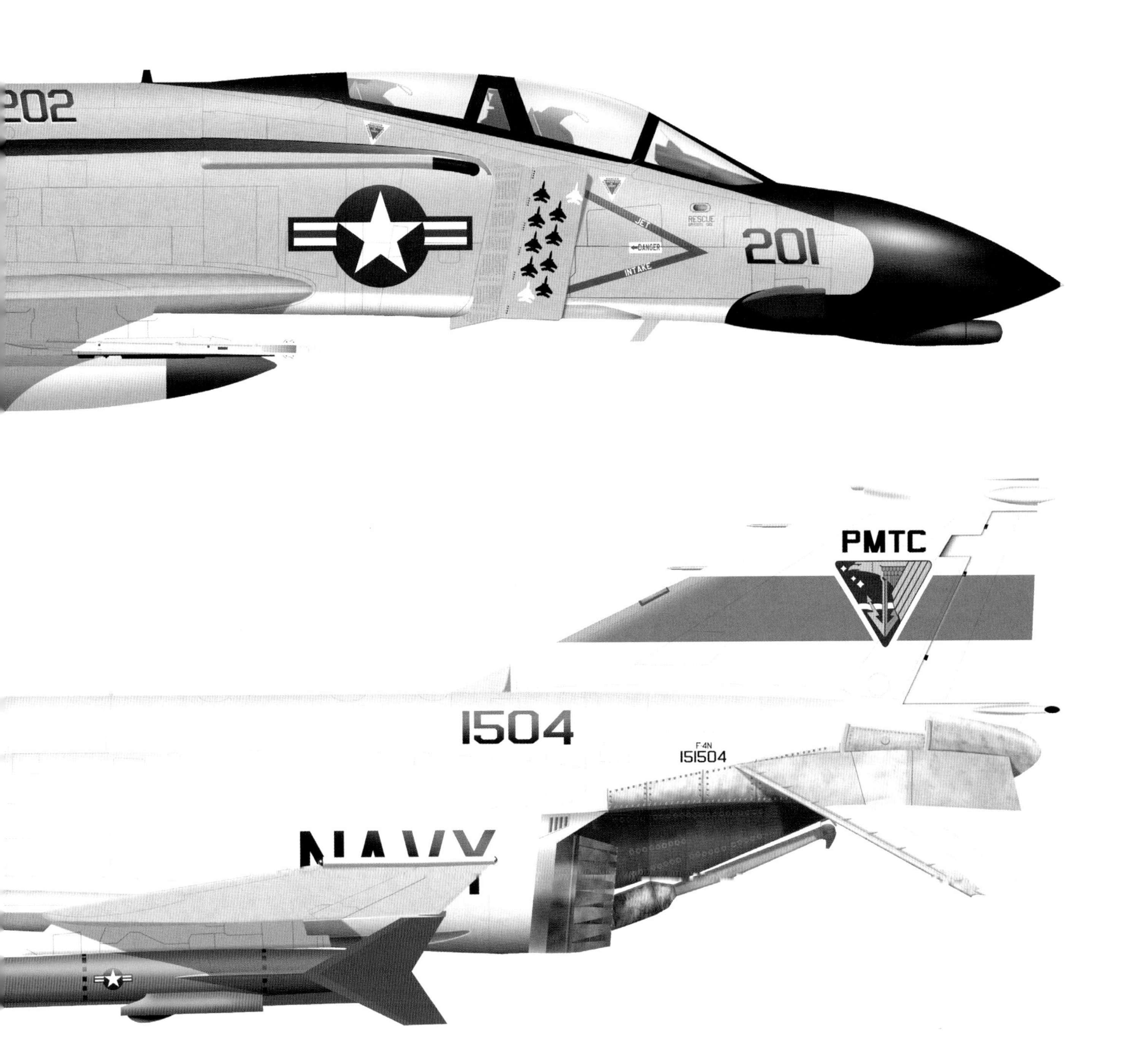

was the inclusion of Martin-Baker zero-zero ejection seats. Internal fuel tank capacity was further increased, mirroring improvements made for the USAF's F-4E, and the main undercarriage wheels were also enlarged in line with those of the Air Force's aircraft.

A new tailplane had a slotted leading edge and a system which drooped the ailerons by 16.5° when the flaps were down to shorten take-off distance and reduce approach speed. The aircraft was also fitted with a new automatic flaps system. A trio of F-4Bs were converted into YF-4Js, with the first one flying on June 4, 1965. Testing proved successful and the first production F-4J flew on May 27, 1966. A total of 522 F-4Js were built with the final aircraft being delivered in January 1972.

A pair of F-4Js were converted into EF-4J electronic aggressor aircraft for VAQ-33 and 15 were modified to accept British electronics systems and supplied to the RAF as the F-4J(UK) in 1984. A single F-4J became a DF-4J drone controller.

During the Vietnam War, US Navy F-4 Phantom squadrons took part in 84 combat tours with F-4Bs, F-4Js and a handful of F-4Ns (see below), claiming a total of 40 air-to-air victories to 73 Phantoms lost in combat – seven to enemy aircraft, 13 to SAMs and 53 to AAA. ▶

McDonnell Douglas F-4 Phantom II

Another 54 Navy Phantoms were lost due to accidents and mechanical failure.

The Blue Angels demonstration team switched to F-4Js on December 23, 1968. The team's seven aircraft each had their AWG-10 weapons control systems removed and replaced with ballast to maintain the aircraft's centre of gravity. The variable intake ramps were disabled, since the team only flew at subsonic speeds, and four dummy Sparrow missiles were attached below the fuselage. The front two contained red and blue dye while the two at the rear contained oil for creating smoke.

The Blue Angels F-4Js also received a fuel tank modification so that they could fly inverted for more than 30 seconds. Their cockpits were altered, with the weapons selector removed and switches added which activated the smoke and dye systems. The automatic flaps system was also removed. Perhaps the biggest difference between the team's F-4Js and the standard model was their engines – they were fitted with the old J79-GE-8s.

Following the introduction of the F-4J, Project Bee Line was launched to bring 228 surviving F-4Bs up to a similar standard. This included an electronics upgrade with AN/ASW-25 datalink, ECM equipment and the F-4J's helmet-mounted sight system. It also saw the

MCDONNELL DOUGLAS F-4S PHANTOM II

McDonnell Douglas F-4S Phantom II, ND-111, VF-301, US Naval Reserve, NAS Miramar, California, 1983. Both the USAF and US Navy experimented with several camouflage schemes designed by artist Keith Ferris. Although successful in tests, the colour schemes would not be adopted into widespread service. VF-301 would paint their F-4 Phantoms in a revised colour scheme known as Heater-Ferris.

MCDONNELL-DOUGLAS F-4S PHANTOM II

McDonnell Douglas F-4S Phantom II, AA-104, VF-74, USS *Forrestal* (CVA-56), 1982. VF-74 became the first squadron to deploy with the Phantom in 1961, adopting the motto 'First in Phantoms'; it operated several variants until 1983. This Phantom shows the Tactical Paint Scheme colours and markings.

installation of Sidewinder Expanded Acquisition Mode (SEAM), IFF, and a dogfight computer. The old J79-GE-8 engines were removed and replaced with a new smokeless version of the J79-GE-10; the airframes themselves underwent structural repairs and strengthening and the aerodynamic improvements of the F-4J such as the slotted tailplanes were added. The 228 aircraft received the new designation F-4N and the first completed example flew on June 4, 1972.

Starting in June 1975, a similar programme of upgrades was applied to the F-4J itself. A total of 265 surviving airframes were strengthened, the old wiring loom was removed and replaced, and wings were modified with leading edge slats to improve manoeuvrability. A new FCS, the AN/AWG-10B, was installed and radios were upgraded. Improvements were made to the avionics cooling vents and the old J79-GE-10s were replaced with the new smokeless version. These 265 aircraft received the designation F-4S.

The remaining F-4Ns were retired in 1984 and the last F-4Ss were retired in 1987. The final Phantoms to see service with the Navy were QF-4N and QF-4S target drones, which were retired in 2004.●

GRUMMAN F-14 TOMCAT

The iconic variable geometry F-14 fleet defence fighter became a symbol of US Navy air power during the 1970s and 80s, serving with distinction well into the 2000s. Heavily armed and flying at speeds above Mach 2 – but still able to land on a carrier deck – it was one of the most capable fighters ever to serve with the Navy.

GRUMMAN F-14A TOMCAT

Grumman F-14A Tomcat, 6, NAS Point Mugu, California, 1972. The AIM-54 Phoenix long-range AAM and the AWG-9 radar constitute the core systems of the Tomcat's offensive capability. Here one of the prototypes is seen during missile separation tests.

Grumman's first experience of 'swing-wing' aircraft design was with the XF10F-1 Jaguar. This troublesome would-be fighter never got past the prototype stage in the early 1950s but it did at least provide the company with a wealth of valuable test data.

And it proved that the principle was sound. If an aircraft's wings could be swept forward, becoming almost straight during take-off and landing, the amount of lift generated would considerably shorten the length of runway required. This would allow heavier payloads to be carried which in turn meant improved range and better armament.

Once airborne, however, if the wings could then be swept sharply back, the aircraft would benefit from much higher speeds than would be possible with straight wings. It was a solution which offered the best of both world, if it could be made to work correctly.

In 1961, US Secretary of Defense Robert McNamara compelled the USAF and US Navy to develop a single aircraft that could meet the requirements of both for a next generation fighter with variable geometry wings. The result was the TFX competition, which Grumman did not enter, with General Dynamics being declared the winner with its F-111 design.

The plan was to create an F-111A which met the Air Force's requirements and an F-111B for the Navy. However, General Dynamics had little knowledge of carrier aircraft design and therefore teamed up with Grumman to assemble and test the 'B' aircraft. Grumman even ended up building the F-111A's rear fuselage and landing gear too.

After seven years of work, the F-111B was cancelled in 1968. However, several years earlier when it was evident that the F-111B was unlikely to become a successful fighter, the US Navy asked Grumman to consider alternatives. The result was a new set of seven concept designs which the company named 303. Some of these had fixed wings, others variable geometry. There were low wings, high wings, engines in the fuselage configurations and arrangements where the engines were in wing-mounted nacelles. After 9000 hours of wind tunnel testing and a total of more than 2000 configurations, a design referred to as the 303E was chosen for further development.

In July 1968, with both the F-111B and McNamara himself now gone, Naval Air Systems Command launched a new programme known as Naval Fighter Experimental (VFX) and asked for proposals from industry. The new fighter had to be a Mach 2.2 tandem two-seater with two engines and armament ▶

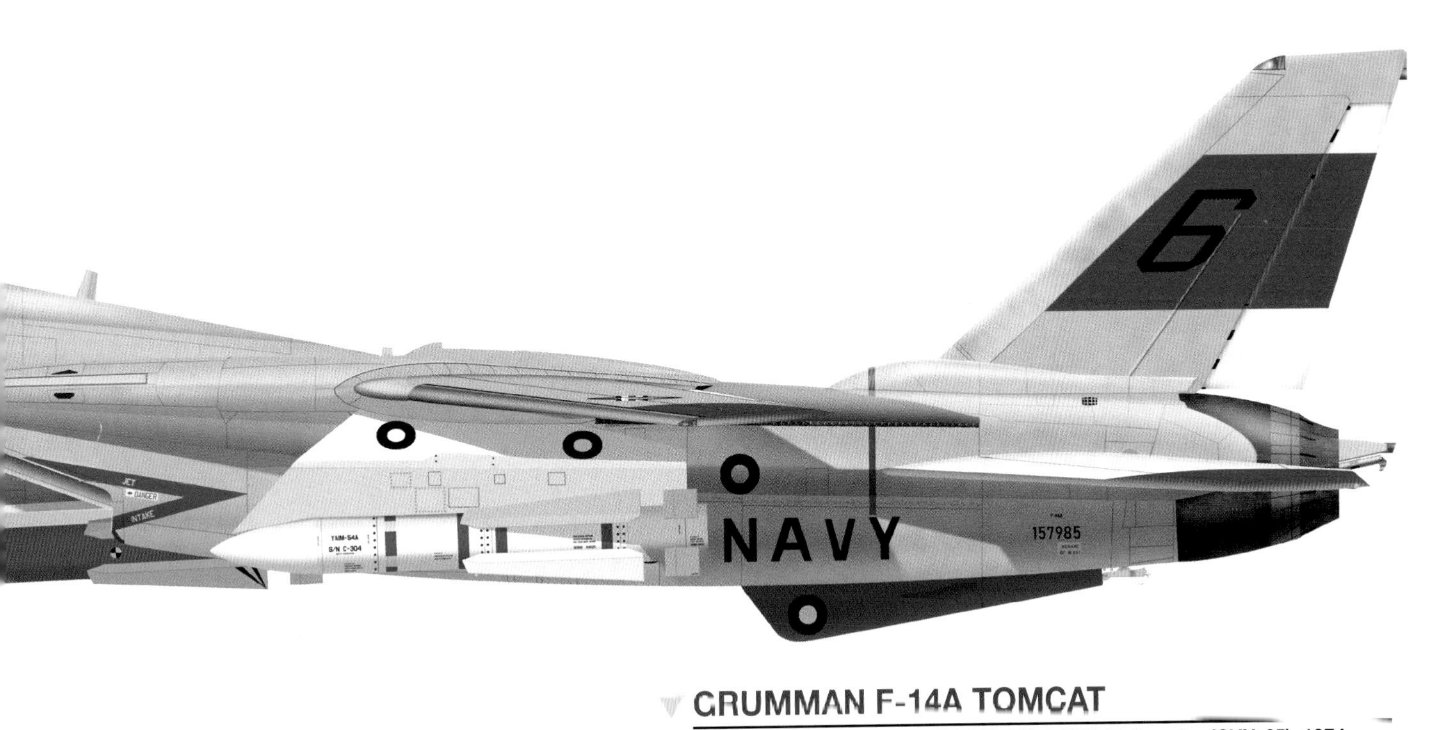

GRUMMAN F-14A TOMCAT

Grumman F-14A Tomcat, NK-103, VF-1, USS *Enterprise* (CVN-65), 1974. The initial operational career of the F-14 coincided with the last phase of US involvement in Vietnam; some flights were made by Tomcats over Saigon (during Operation Frequent Wind) but no engagement was recorded.

GRUMMAN F-14A TOMCAT

Grumman F-14A Tomcat, 226, PMTC, NAS Point Mugu, California, 1981. Although the AIM-120 AMRAAM was tested with Tomcats, the missile was never integrated into operational aircraft.

GRUMMAN F-14A TOMCAT

Grumman F-14A Tomcat, 102, VF-41, USS *Nimitz* (CVN-68), 1981. The US Navy combat debut of the F-14 happened over Libya in 1981; On August 19, 1981, two F-14s from VF-41 shot down two Libyan Su-22s using AIM-9 Sidewinder AAMs.

GRUMMAN F-14A TOMCAT

Grumman F-14A Tomcat, 202, NAWC, NAS Patuxent River, Maryland, 1994. From early on in the Tomcat programme, consideration was given to providing the aircraft with a ground-attack capability; here one of Naval Air Weapons Center's Tomcats is seen carrying laser guided bombs which would, eventually, be deployed with operational F-14s.

Grumman F-14 Tomcat

GRUMMAN F-14A TOMCAT

Grumman F-14A Tomcat, NE-103, VF-1, USS *Ranger* (CV-61), 1991. During Operation Desert Storm, F-14s were assigned mostly to fleet defence and escort missions. This resulted in fewer engagements with enemy fighters and the F-14 ended the war with only one recorded air-to-air kill (an Mi-8 helicopter with an AIM-9 Sidewinder AAM). Curiously the kill mark is a silhouette of an Mi-24 instead of an Mi-8.

GRUMMAN F-14A TOMCAT

Grumman F-14A Tomcat, NJ-76, VF-124, NAS Miramar, California,1976. To celebrate the 200th anniversary of the American Revolution, several US aircraft were painted in commemorative schemes. The F-14 was no exception and this VF-124 Tomcat is visually very striking.

of a single 20mm cannon and either six Hughes AIM-54 Phoenix missiles or a mix of AIM-54s, AIM-7 Sparrows and AIM-9 Sidewinders.

Five companies offered designs, with McDonnell Douglas and Grumman being quickly chosen as the finalists. The contract went to Grumman's 303E in January 1969. Having been burned by the F-111B, the Navy opted to skip the prototype phase in order to avoid political interference and go straight to the first production version of what was now designated the F-14.

The F-14A's cockpit was a tandem two-seater, as per the spec, with both pilot and radar intercept officer sitting on Martin-Baker GRU-7A zero-zero ejection seats beneath a clamshell canopy. On the right side of the aircraft's nose was a retractable inflight refuelling probe.

Behind the cockpit was a broad flat fuselage with the wings attached to non-moving 'gloves' which housed their sweep mechanism. They also housed triangular 'vanes' in their leading edge which popped out at supersonic speeds to improve stability. Beneath the gloves and widely spaced on either side were the aircraft's two Pratt & Whitney TF30-P-412 engines – with 12,350lb dry thrust or 20,900lb with afterburner – in what amounted to under-slung nacelles. They were fed by wedge-shaped intakes on either side of the fuselage, blended into the gloves.

The wing sweep mechanism was linked to a system which automatically varied their position between 20° and 68° depending on the speed and attitude of the aircraft, although the pilot could override this and set the angle manually.

When the aircraft was in storage, the wings could be set to a 75° sweep which overlapped them with the all-moving tailplanes to save space. At the rear of the aircraft were two vertical fins and underneath the rear section of each engine housing was a short ventral fin.

The undercarriage mainwheels were housed in the underside of the wing gloves and retracted forwards, as did the nosewheel. Armament was nominally, as specified, a General Electric M61A1 Vulcan 20mm cannon under the left hand side of the cockpit plus six AIM-54s – although these enormous missiles were so heavy that the F-14 was unable to land on a carrier with six of them. Four were attached to the aircraft's fuselage between its engines and two were on wing glove pylons. However, in service the aircraft was typically equipped with the four fuselage AIM-54s plus two AIM-7s and two AIM-9s – the latter four on the wing glove pylons.

The AIM-54s and AIM-7s were linked to a Hughes AN/AWG-9 radar with AN/AWG-15 fire control system, managed by the radar intercept officer in the back seat. Under its nose, the F-14A had an AN/ALR-23 Infrared Search and Track sensor. Other key systems included AN/ALE-39 chaff and flare dispensers, an AN/APR-45 radar warning receiver and an AN/ALQ-126 electronic jamming system.

Structurally, the aircraft was primarily made from lightweight aluminium alloy but around 25% was titanium, including the wing box, wing pivots, upper and lower wing skins. The Phantom II had made limited use of the material, around 9% of its structure, but the F-14's far ▶

greater percentage ensured that the aircraft had the optimum strength to weight ratio.

The first operational units to receive the F-14A were VF-1 and VF-2 on board USS *Enterprise* in September 1974 and the type gradually replaced the Navy's F-4 Phantoms and F-8 Crusaders in the fleet defence role. Tomcats would also replace many of the Navy's other reconnaissance types, such as the RF-8G Crusader, thanks to the Tactical Air Reconnaissance Pod System or 'TARPS'.

Development of this piece of equipment, housed within an aerodynamic pod measuring 5.18m long, commenced in 1979. It included a Honeywell AN/AAD-5 infrared line scanner, a Fairchild KA-99 panoramic camera and a CAI KS-87B serial frame camera. It was carried beneath the fuselage in the rear fuselage tunnel and since it required its own power supply, not to mention a TARPS display for the radar intercept officer, special modifications were required before an F-14A could carry it. About 50 aircraft received these upgrades starting in 1980 and TARPS Tomcats became the Navy's main reconnaissance asset going into the early 2000s.

F-14B, C AND D

During the early years of the F-14A's service the TF30-P-412 engines suffered from reliability issues – partly because the extreme manoeuvrability and performance of the aircraft subjected them to previously unheard-of stresses during flight. Fan blades from the power plant could break loose, cutting their way

through the fuselage at high speed. The engines were so widely spaced that if one failed while the afterburners were lit the asymmetrical thrust could cause the aircraft to enter a spin that could not be recovered.

As a result, all F-14As had been fitted with the improved TF30-P-414 by 1979 and further upgrades were made in 1981, resulting in the TF30-P-414A.

Pratt & Whitney had originally intended the TF30 as an interim engine while its F401-P-400 afterburning turbofan – the Navy's version of the successful F100 (which still powers the F-15 today) was readied for series production. It was projected that F-14A production would be switched to the new engine with the designation F-14B as soon as it became available. The seventh pre-production F-14 was fitted with F401s and made its first flight in this configuration on September 12, 1973, but the Navy decided to cancel the engine due to cutbacks and the F-14B was stillborn.

A further development of the F-14 with the TF30-P-414A featuring updated and enhanced avionics, designated F-14C, was also considered but ultimately also cancelled. Circumstances had changed slightly by the early 80s, however, and the Navy still wanted better engines for its Tomcats.

At this time General Electric was hoping to sell a smaller 'fighter' variant of its F101 turbofan – developed to power the Rockwell B-1 Lancer – known as the F101-DFE or 'Derivative Fighter Engine'. The sole F-14B development mule was refitted with two prototypes of this powerplant and tested – with overwhelmingly positive results.

GRUMMAN F-14A TOMCAT

Grumman F-14A Tomcat, 13, NSAWC, NAS Fallon, Nevada, 1997. Topgun is the popular name of the United States Navy Strike Fighter Tactics Instructor programme (SFTI programme) in the US Navy Fighter Weapons School. Aircraft were used as adversaries to give operational crews a realistic training experience. In 1996 this school was merged with the Naval Strike Warfare Center and the Carrier Airborne Early Warning Weapons School to form the Naval Strike & Air Warfare Center (NSAWC). It used F-14s until 2003.

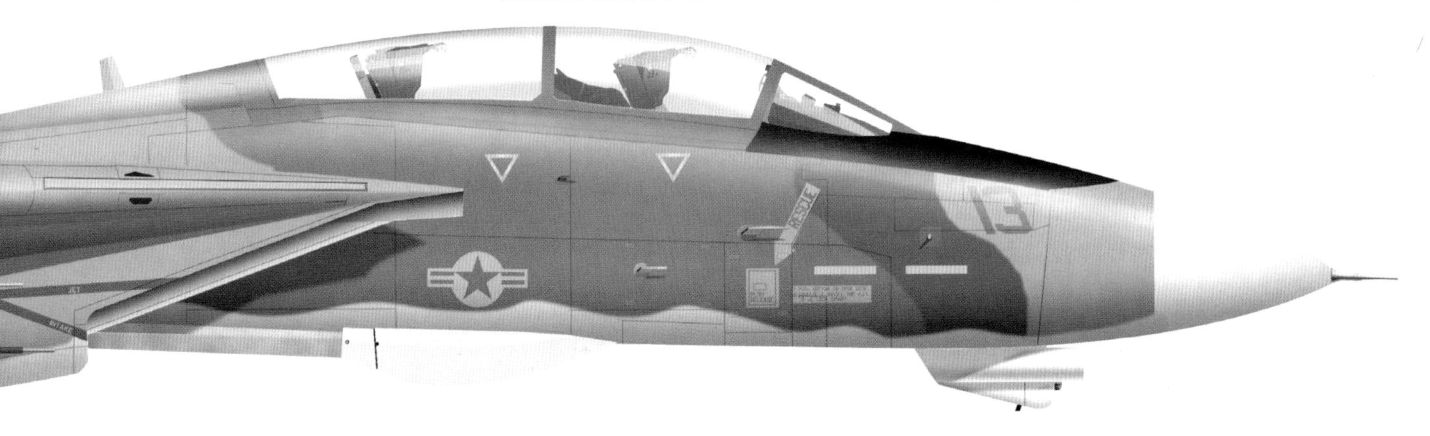

GRUMMAN F-14B TOMCAT

Grumman F-14B Tomcat, AG-103, VF-143, USS *Dwight D. Eisenhower* (CVN-69), 1991. The F14B variant introduced several improvements, among those the long-awaited new engine. VF-143 was deployed during Operation Desert Shield and saw combat in the subsequent Operation Desert Storm.

Grumman F-14 Tomcat

GRUMMAN F-14B TOMCAT

Grumman F-14B Tomcat, AC-104, VF-32, USS *Harry S. Truman* (CVN-75), 2003. Air-to-ground weapons became a more common sight in the latter years of the aircraft's career with the US Navy. Here a VF-32 aircraft carries impressive air-to-ground mission markings. VF-32 was deployed during Operation Iraqi Freedom.

GRUMMAN F-14B TOMCAT

Grumman F-14B Tomcat, AA-105, VF-74, NAS Oceana, Virginia, 1993. VF-74 painted their Tomcats in this colour scheme for their role as adversaries.

GRUMMAN F-14D TOMCAT

Grumman F-14D Tomcat, NK-107, VF-31, USS *Abraham Lincoln* (CVN-72), 1998. The F-14D variant featured a new engine (F-110-GE-400), a new radar (AN/APG-71) and a glass cockpit. Other improvements included new ejection seats and upgraded avionics. VF-31 alongside VF-213 would be the last US Navy units to deploy with the F-14.

Grumman F-14 Tomcat

GRUMMAN F-14A TOMCAT

Grumman F-14A Tomcat, AF-101, VF-201, US Naval Reserve, NASJRB Fort Worth, Texas, 1999. The F-14 was also operated by US Naval Reserve (USNR) squadrons and US Naval Reserve Augmentation Units (USNRAU). VF-201 operated F-14s until 1999.

GRUMMAN F-14D TOMCAT

Grumman F-14D Tomcat, NE-100, VF-2, USS *Constellation* (CV-64), 2003. Contrary to the general trend in military aviation and following US Navy tradition, the 'CAG Birds' (the official Air group's Commander aircraft), usually displayed less subdue looks, as can be seen here on this VF-2 aircraft. It also carries the TARPS reconnaissance pod.

The Navy therefore commissioned General Electric to refine the F101-DFE into a full production variant under the new designation F110. The F-14B was flown with F110-GE-400 engines for the first time on September 20, 1986, and a further five F-14As were also refitted to undertake testing and further development work. Crucially, the F110 was very similar in both size and shape to the TF30, which allowed for almost a straight swap with only minor modifications.

This progressed rapidly and a contract for production model F110 engines was placed on February 15, 1987. A single one of these engines produced 13,800lb of dry thrust and 23,400lb with afterburner – representing a very healthy increase of 1450lb and 2500lb respectively compared to the TF30. The extra power allowed the F-14 to take off from a carrier without using afterburner, which saved fuel and therefore increased range, and the engines were much more easily able to keep up with the aircraft's demands during extreme manoeuvring too.

Grumman was contracted to build 38 new Tomcats fitted with the F110 as standard under the designation F-14A (Plus) as well as bringing 32 of the original F-14As up to F-14A (Plus) standard. Visually, the improved aircraft were very similar to the original Tomcats – except for larger exhaust nozzles, antenna for the AN/ALR-67 radar warning receiver added beneath the wing gloves and a detail change

to an access hatch close to the gun port. Perhaps the biggest giveaway was the deletion of the glove vanes. These aircraft began to enter service towards the end of 1988 and three years later the F110-powered F-14As were redesignated F-14B.

In parallel to this, the Navy also ordered a thoroughly modernised new Tomcat. This variant would be powered by F110s like the 'B' but it also had a full avionics upgrade. The cockpit layout was significantly improved to take account of experience with the F-14A, new displays were installed that worked with night-vision goggles, the latest Martin-Baker ejection seats were installed, and the new AN/APG-71 radar system – based on the F-15E's AN/APG-70 – was fitted. InfraRed Search and Track and Television Camera Set sensors were mounted in a 'double barrel' chin pod that clearly distinguished the new F-14 from its older siblings.

This time only four F-14As were converted as development aircraft and the first flew with its new engines and systems on November 24, 1987. The upgrade was a great success and the new variant was designated F-14D. Yet despite the significantly increased capabilities of the F-14D, only 37 brand new examples were purchased with the first of them entering squadron service during November 1990. A further 18 F-14As were brought up to 'D' standard, being designated F-14D(R).

In total, an incredible 557 F-14As were built, plus the 38 new F-14Bs and 37 new ▶

Grumman F-14 Tomcat

F-14Ds for a grand total of 632 US Navy Tomcats. A further 80 F-14As were built for Iran but that is another story which falls beyond the remit of this publication.

In addition, in 1988, the Navy began a modernisation programme for its older F-14As, known as MMCAP for 'Multi-Mission Capability Avionics Program'. This involved the installation of new mission computers, chaff-flare dispensers on the rear of the Sidewinder rails, programmable displays for the radar intercept officer and fitment of the new AN/ALR-67 radar warning receiver.

Deliveries of the upgraded aircraft commenced in 1994 and at the same time work began on a new Digital Flight Control System or DFCS to replace the F-14's original analogue controls. The DFCS included automated flight control computer commands which intervened to minimise dutch roll characteristics, prevent spins and make landing easier.

Development on the new system commenced in 1995 and installations on service aircraft commenced in 1998. Final installations were carried out in early 2001. Meanwhile, the TARPS pod was upgraded to include a digital camera in 1996 and a datalink to allow transmission of images was fitted in 1999.

GROUND-ATTACK

The F-14A had been designed for pure fleet defence – entirely lacking in any form of ground-attack capability, which was unusual for a Navy fighter. However, as funding grew tight during the 1980s it was decided that the Tomcat ought to have greater multirole capability.

Therefore, in 1987, efforts were made to see whether the aircraft could be made to carry bombs.

Eventually, by 1992, the F-14 had been cleared to carry basic 'iron bombs'. Nothing more advanced could be equipped since the aircraft lacked the necessary means to designate ground targets.

Plans were drawn up for another comprehensive upgrade which would see the Tomcat fitted with the appropriate electronic systems and wiring for true ground-attack capability but the cost proved too great. A more affordable solution presented itself in the form of Lockheed Martin's Low Altitude Navigation and Targeting Infra-Red for Night (LANTIRN) targeting pod. With this installed, an F-14 would have both a target designator for laser-guided bombs and a forward-looking infra-red camera for operating at night. The first Tomcat equipped with a LANTIRN pod flew on March 21, 1995, and the system officially entered service on June 14, 1996 – with the aircraft being nicknamed the 'Bombcat'.

The last Tomcat carrier operations took place on July 28, 2006, with the official final flight of a US Navy F-14 taking place on September 22 of the same year. The actual last flight of an F-14 in US service, however, was 12 days later – on October 4.

The following year, the US Navy announced plans to shred the remaining serviceable F-14s to prevent Iran from obtaining any components to keep their fleet flying. This work had been carried out by 2009 – a truly sad end to one of the greatest naval jet fighters in history.●

GRUMMAN F-14D TOMCAT

Grumman F-14D Tomcat, XF-1, VX-9, NAS Point Mugu, California. F-14s were operated by test units of the US Navy. Air Test and Evaluation Squadron Nine (VX-9) was created by the merger of VX-4 and VX-5 in 1993; it operated Tomcats until 2006.

GRUMMAN F-14D TOMCAT

Grumman F-14D Tomcat, AJ-204, VF-213, USS *Theodore Roosevelt* (CVN-71), 2006. VF-213 operated Tomcats longer than almost any other unit, 1976 until 2006. The aircraft received a special paint scheme for the last cruise of the F-14 in 2006.

MCDONNELL DOUGLAS F/A-18A/B/C/D

Designed from the beginning as a lightweight cost-effective multirole fighter, the original F/A-18 proved to be highly versatile, thoroughly reliable and remarkably capable. It continues in service with the USMC but was retired by the Navy in 2018.

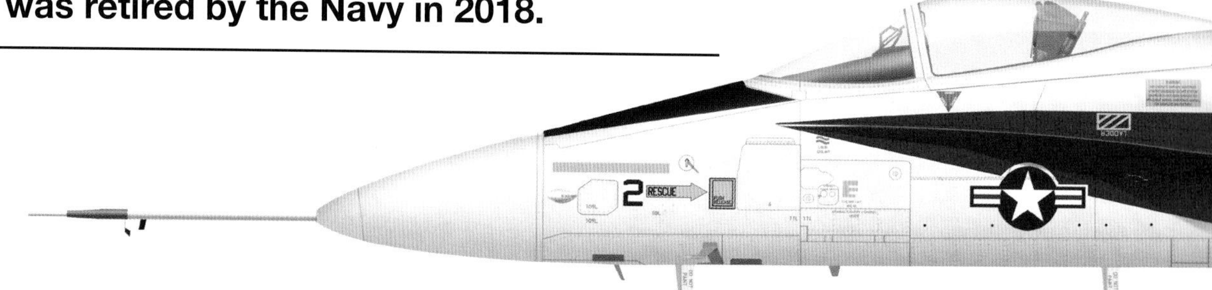

1984-2018

MCDONNELL DOUGLAS F/A-18A HORNET

McDonnell Douglas F/A-18A Hornet, NK-411, VFA-25, USS *Constellation* (CV-64), 1985. The first squadrons to receive the new Hornets were VFA-25 and VFA-113 in 1983; both squadrons embarked for the first Hornet cruise aboard USS *Constellation* in 1985.

Northrop began working on a simple lightweight supersonic fighter in 1952 and eventually produced the F-5 – which would become a huge hit on the international market. This prompted the company to consider plans for a replacement, starting with basic concepts in 1966 and progressing to a much more polished design by 1970.

The aircraft on Northrop's drawing board was the result of extensive wind tunnel testing and experimentation with the form of the original F-5. It still retained the character of an F-5 but with substantial Leading Edge Root eXtensions (LERX) to its wings for improved handling at high angles of attack and two tailfins rather than the original single fin. The latter feature would improve manoeuvrability and enhance stability.

A full scale mock-up of the design, known as the P-530 or 'Cobra' because its LERX gave it a 'hooded' appearance, was displayed at the 1971 Paris Air Show in order to attract interest from the F-5's customers and both single- and twin-engine variants were worked on as the P-610 and P-600 respectively.

The following year, with interest in the idea of a highly manoeuvrable lightweight fighter (dubbed 'LWF') to complement the heavyweight F-15 gaining traction, the USAF launched a new competition for LWF designs on January 6, 1972. Proposals were put forward by Boeing, General Dynamics and Ling-Temco-Vought, with Northrop submitting the P-600.

The USAF ordered two prototypes from both General Dynamics and Northrop for a 'fly-off' contest with the former receiving the designation YF-16 and the latter receiving YF-17. The Air Force set an Air Combat Fighter (ACF) specification for the types in April 1974 and the first YF-17 flew on June 9, 1974, being joined by the second on August 21 of the same year. Each was powered by a pair of YJ101-GE-100 turbofans producing 14,400lb of thrust with afterburner. However, the YF-16 was chosen as the ACF by a close margin, becoming the famous F-16.

Meanwhile, the Navy had been pushed to seek its own LWF which could complement the heavier F-14 as well as replacing A-7s and F-4s in the strike-fighter role. This resulted in what became the Navy ACF or 'NACF' competition, with General Dynamics joining forces with Ling-Temco-Vought to create a navalised YF-16 and McDonnell Douglas joining Northrop to navalise the YF-17.

The Navy announced on May 2, 1975, that the YF-17 was the winner and would ▶

MCDONNELL DOUGLAS F/A-18A HORNET

McDonnell Douglas F/A-18A Hornet, 2, NAS Patuxent River, Maryland, 1981. One of the Full Scale Development aircraft is seen here during the test programme. Destined for both services, the aircraft displayed NAVY on the aft lower port fuselage and MARINES on the other side.

McDonnell Douglas F/A-18A/B/C/D

MCDONNELL DOUGLAS/BOEING F/A-18A+ HORNET

McDonnell Douglas/Boeing F/A-18A+ Hornet, X-01, VFA-204, Naval Air Station Joint Reserve Base New Orleans, Louisiana, 2011. VFA-204 is a US Naval Reserve Squadron, assigned primarily for operational and training support of active forces, providing adversary training. One of its aircraft was painted in this retro colour scheme to commemorate the 100th anniversary of US naval aviation.

MCDONNELL DOUGLAS F/A-18A HORNET

McDonnell Douglas F/A-18A Hornet, AK-211. VFA-132, USS *Coral Sea* (CV-43), 1986. F/A-18s made their combat debut in 1986 during Operation Prairie Fire, part of Operation El Dorado Canyon, the bombing of Libyan targets. Hornets performed Suppression of Enemy Air Defense (SEAD) missions.

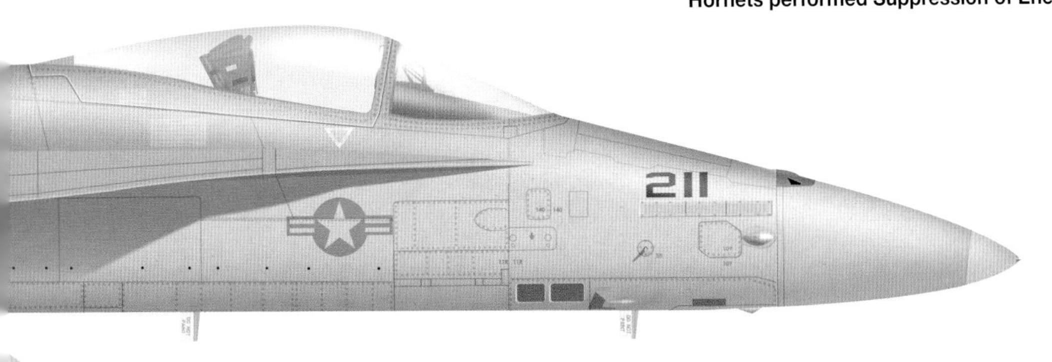

MCDONNELL DOUGLAS/BOEING F/A-18A HORNET

McDonnell Douglas/Boeing F/A-18A Hornet, 5, Blue Angels, 2008. The Blue Angels received Hornets in 1986 and continue to use them to this day, having gone though several variants (A, C and now starting to receive the new E Super Hornets). The team have used the F/A-18 Hornet longer than any other aircraft (35 years).

McDonnell Douglas F/A-18A/B/C/D

be built as the F-18. McDonnell Douglas and Northrop signed an agreement that MD would be the lead constructor of the naval F-18 while Northrop would lead on the land version, the F-18L – with which they still hoped to attract many international customers.

Unfortunately for Northrop, the F-18L found no takers and the aircraft they had designed and developed was now being sold to the US Navy with McDonnell Douglas as the manufacturer. The Navy ordered a total of 11 F-18s – a pair of two-seaters and nine single seaters – for evaluation and McDonnell Douglas duly set about converting the YF-17 for carrier operations.

This involved the installation of a more powerful development of the YJ101 – the General Electric F404 – bigger wings that could also be folded, high-lift devices for deck landings, stronger landing gear, increased fuel capacity and in-flight refuelling capability. All of this added a considerable amount of weight to the original YF-17 design and made the F-18 Hornet, as it was now named, almost a heavyweight in its own right.

The first F-18 prototype flew on November 18, 1978 and the last of the initial 11 examples was received by the Navy in March 1980. These were assigned to VFA-125, the Fleet Replacement Squadron for training flight crews. The testing and evaluation programme which followed lasted until October 1982, by which time the type had been redesignated F/A-18 as an indicator of its multirole capability.

MCDONNELL DOUGLAS/BOEING F/A-18B HORNET

McDonnell Douglas/Boeing F/A-18B Hornet, AF-12, VFC-12, NAS Oceana, Virginia, 2017. VFC-12 is a composite squadron of the US Naval Reserve, whose primary mission is providing training for operational F/A-18 squadrons. This aircraft was painted in a camouflage scheme similar to one used on Sukhoi Su-34s.

MCDONNELL DOUGLAS/BOEING F/A-18C HORNET

McDonnell Douglas/Boeing F/A-18C Hornet, AA-410, VFA-81, USS *Saratoga* (CV-60), 1991. Two aircraft from VFA-81 achieved the first US Navy kills of Operation Desert Storm (two MiG-21s) on the first day of the air campaign.

Structurally, the production model F/A-18A Hornet was 50% aluminium, 17% steel, 13% titanium and 10% glass reinforced plastic. It was powered by two F404-GE-400s, each providing 16,000lb-ft of thrust with afterburner, with D-shaped intakes and splitter plates. Its wings were about 14% larger than those of the YF-17 and featured a power-folding mechanism.

The Hornet also had a strong undercarriage with two wheels on the nose and a catapult attachment included. An arresting hook was installed to the rear of the aircraft.

Inside the cockpit, beneath a full-vision canopy, the pilot sat on a Martin-Baker Mark 10 ejection seat and faced three multifunction displays linked to the aircraft's Hughes AN/APG-65 radar, which could be toggled between navigation, air combat and strike modes. The aircraft also featured an AN/ALQ-126B jamming system, AN/ALE-39 chaff and flare dispensers and an AN/ALR-50 radar warning system, plus the usual radio, IFF and navigation beacon.

Built-in armament was a single M61A1 Vulcan 20mm cannon mounted centrally in the top of the nose. This position shielded the pilot from muzzle flash – preventing night blindness – and allowed gases from the weapon to pass over the LERX without entering the engine intakes. A special mounting system was included to prevent vibrations from the cannon affecting the delicate electronic instruments installed close by. ▶

McDonnell Douglas F/A-18A/B/C/D

Each wingtip had a single Sidewinder launch rail and there were two pylons for external stores beneath each wing. In addition, a single Sparrow-type missile could be recessed into fuselage on either side and there was a centreline fuselage hardpoint too.

Maximum external load was 15,500lb and a huge range of stores could be accommodated including Zuni rocket pods, AGM-88 HARM anti-radar missiles, AGM-65 Maverick air-to-surface missiles, AN/AAS-38 Nite Hawk targeting pods, jamming pods, 1250 litre external fuel tanks and much more. Using external tanks and the buddy system, an F/A-18A could even operate as a tanker.

McDonnell Douglas manufactured a total of 371 F/A-18As between 1980 and 1987.

F/A-18B/C/D

The F/A-18B was the two-seat variant with a second cockpit behind the first and dual controls. The aircraft itself was nearly identical to the F/A-18 in every other respect – the extra cockpit space being created by sacrificing about 6% of internal fuel capacity and moving some of the electronics around. Although the two-seater was fully combat capable, the 40 examples built nearly all served as operational conversion trainers.

There was a brief attempt to create a dedicated reconnaissance Hornet by adding a camera nose to a single F/A-18B airframe but although the work was completed and the aircraft flew on August 15, 1984, no production order was forthcoming. The USMC would

MCDONNELL DOUGLAS/BOEING F/A-18C HORNET

McDonnell Douglas/Boeing F/A-18C Hornet, XE-300, VX-9, Naval Air Weapons Station China Lake, California, 2002. VX-9 tests operational weapons, electronic warfare and software. It uses different types of aircraft including legacy Hornets and Super Hornets.

later order F/A-18Ds with nose-mounted reconnaissance packages but the US Navy did not.

During 1987, McDonnell Douglas replaced the single-seat F/A-18A and two-seat F/A-18B on its production line with the single-seat F/A-18C and two-seat F/A-18D. Visually there was little change, except for the addition of a single strake to the rear of each LERX. These were retrofitted to all F/A-18As and Bs so it remained very difficult to tell the newer aircraft apart from their older siblings.

Internally, the F/A-18C differed from the F/A-18A in having improved Martin-Baker Mark 15 ejection seats and electronics upgrades which allowed the aircraft to carry newer versions of the Maverick, the AIM-120 AMRAAM, and the latest Sparrow. It also got the new AN/ALR-67 radar, AN/ALE-47 chaff and flare dispensers, more modern computer systems and an upgraded air flow system to keep them cool.

The F/A-18D was essentially a two-seat F/A-18C – replicating the original F/A-18A and B format. The F/A-18C prototype – a modified F/A-18A – first flew on September 3, 1986.

LATER UPGRADES

Next up on the McDonnell Douglas production line were the F/A-18C+ and D+. This upgrade created the Night Attack Hornet with night vision goggle compatibility and a digital moving map. The first F/A-18C+ was delivered on November 1, 1989, and from 1991 the aircraft could also carry the AN/AAS-38A ▶

MCDONNELL DOUGLAS/BOEING F/A-18C HORNET

McDonnell Douglas/Boeing F/A-18C Hornet, NK-300, VFA-113, USS *Ronald Reagan* (CVN 76), 2009. VFA-113 was deployed to support Operation Enduring Freedom, performing combat operations over Afghanistan in 2009.

MCDONNELL DOUGLAS/BOEING F/A-18C HORNET

McDonnell Douglas/Boeing F/A-18C Hornet, NE-400, VFA-34, USS *Carl Vinson* (CVN-70), 2018. This appropriately painted CAG bird displays a special marking celebrating the last RIMPAC exercise the legacy Hornets participated on.

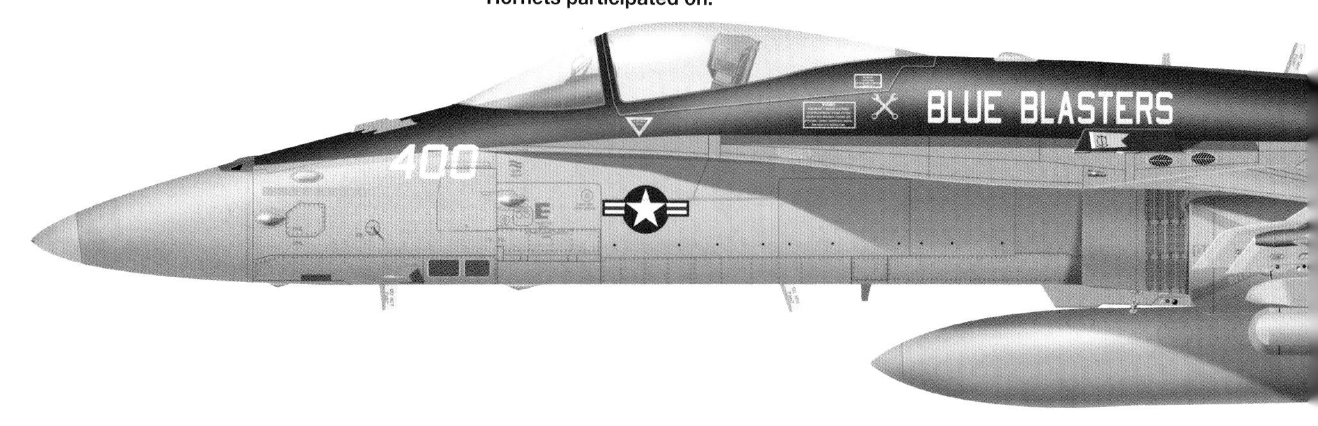

MCDONNELL DOUGLAS/BOEING F/A-18D HORNET

McDonnell Douglas/ Boeing F/A-18D Hornet. 00, USNTPS, NAS Patuxent River, Maryland, 2005. The United States Naval Test Pilot School is one of the last US Navy units to fly legacy Hornets.

MCDONNELL DOUGLAS/BOEING F/A-18C HORNET

McDonnell Douglas/ Boeing F/A-18C Hornet, AD-300, VFA-106, NAS Oceana, Virginia, 2019. The last official active-duty flight of a legacy Hornet of the US Navy was made in October of 2019.

laser targeting pod, followed by the AN/AAS-38B pod in 1996 which had an 'auto-tracking' ability.

One of the most significant upgrades in the Hornet's history took place in 1992 when the fleet was re-engined with the F404-GE-402, developing 17,600lb-ft of thrust with afterburner, and adding increased reliability. Two years later, the aircraft were fitted with the new AN/APG-73 multimode radar which provided increased range and was significantly easier and more intuitive to operate. And a GPS receiver was installed in 1995. Radar absorbing material was also added to provide a degree of stealth capability.

A total of 466 F/A-18Cs and 161 F/A-18Ds were built for the US Navy and USMC. Including export models, a grand total of 1480 F/A-18A/B/C/D variants were built.

Another wave of upgrades to F/A-18s with sufficiently low flight hours commenced in 2000, bringing with it modifications to support the latest AMRAAM, but the days of the US Navy's Hornet fleet were numbered. While the Marine Corps continued to update its fleet, with plans to keep it in service until at least 2030, the Navy began retiring its Hornets in 2017. The last operational deployment was made in late 2018. Some 136 examples have been moved to the Davis-Monthan Air Force Base boneyard in Arizona while examples in a better state have been handed over to the Marine Corps.

Since then, the US Navy has switched to the more advanced Super Hornet.●

BOEING F/A-18E/F SUPER HORNET

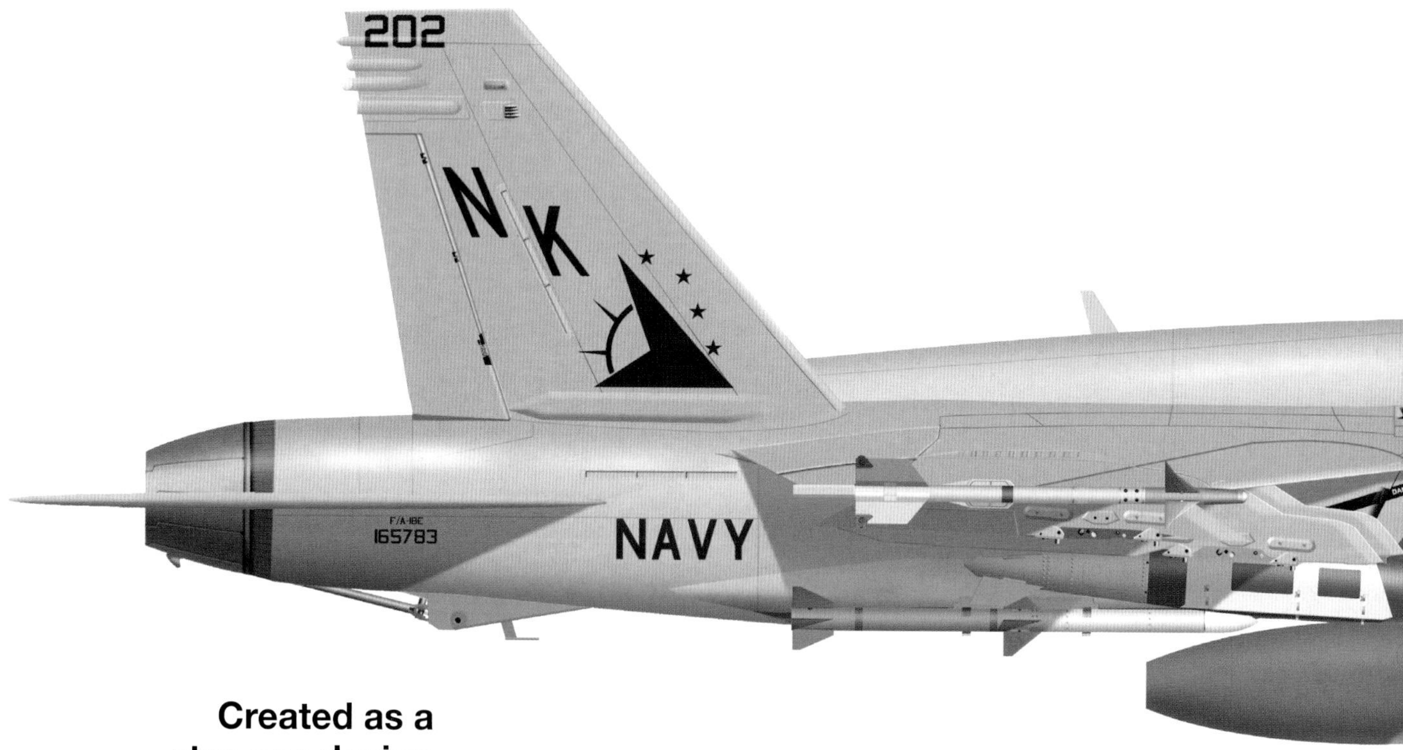

1999-PRESENT

Created as a stopgap design and looking like a scaled-up F/A-18C, the 'Rhino' is a jack-of-all-trades. While some question the benefits of relying on such a generalist, the F/A-18E/F has nevertheless become the mainstay of the US Navy's fighter fleet.

BOEING F/A-18 SUPER HORNET

Boeing F/A-18E Super Hornet, NJ-101, VFA-122, NAS Fallon, Nevada, 2012. VFA-122 is a Fleet replacement squadron, tasked with training Navy and Marine Corps crews for Hornet and Super Hornet squadrons. For some time the squadron's aircraft were painted in this striking colour scheme.

The US Navy had the new F-14 for pure fleet defence going into the early 1980s but lacked a successor to its aging strike force of A-6s, A-7s and F-4s. The multirole F/A-18 was introduced in 1984 but many felt that its capabilities overlapped more with those of the F-14 than the ground-attackers.

Nevertheless, the Reagan administration (1981-1989) was providing lavish funding for defence so a multitude of different Navy attacker programmes were launched. Initially the most promising of these was a significant upgrade of Grumman's A-6 to A-6F standard, with completely new avionics, much more powerful new engines and a consequently increased load-carrying capacity.

However, despite five development aircraft being created, the Navy decided to have a completely new aircraft built instead – the radical delta flying wing McDonnell Douglas A-12 Avenger II. The A-12 programme soon ran into difficulties however, falling behind schedule as costs spiralled, and was cancelled in 1991.

The Navy was by now suffering a serious shortfall in strike capability and during Operation Desert Storm, following Iraq's invasion of Kuwait, contributed 80% fewer sorties than the Air Force. A new strike aircraft was still planned under the designation 'AX' but an interim solution was needed. Consequently, attention focused on the what might be done with the existing F/A-18 Hornet.

It was considered that a thoroughly improved Hornet, building on the reliable and combat-tested aircraft design, would be able to perform both the F-14's fleet defence job and fill in on ground-attack until the AX arrived. McDonnell Douglas received a contract for seven prototypes ▶

BOEING F/A-18E SUPER HORNET

Boeing F/A-18E Super Hornet, NK-202, VFA-115, USS *Abraham Lincoln* (CVN-72), 2002. The Super Hornet achieved initial operational capability in 2001 with VFA-115. This squadron was also the first to use the new aircraft in combat over Iraq, in 2002.

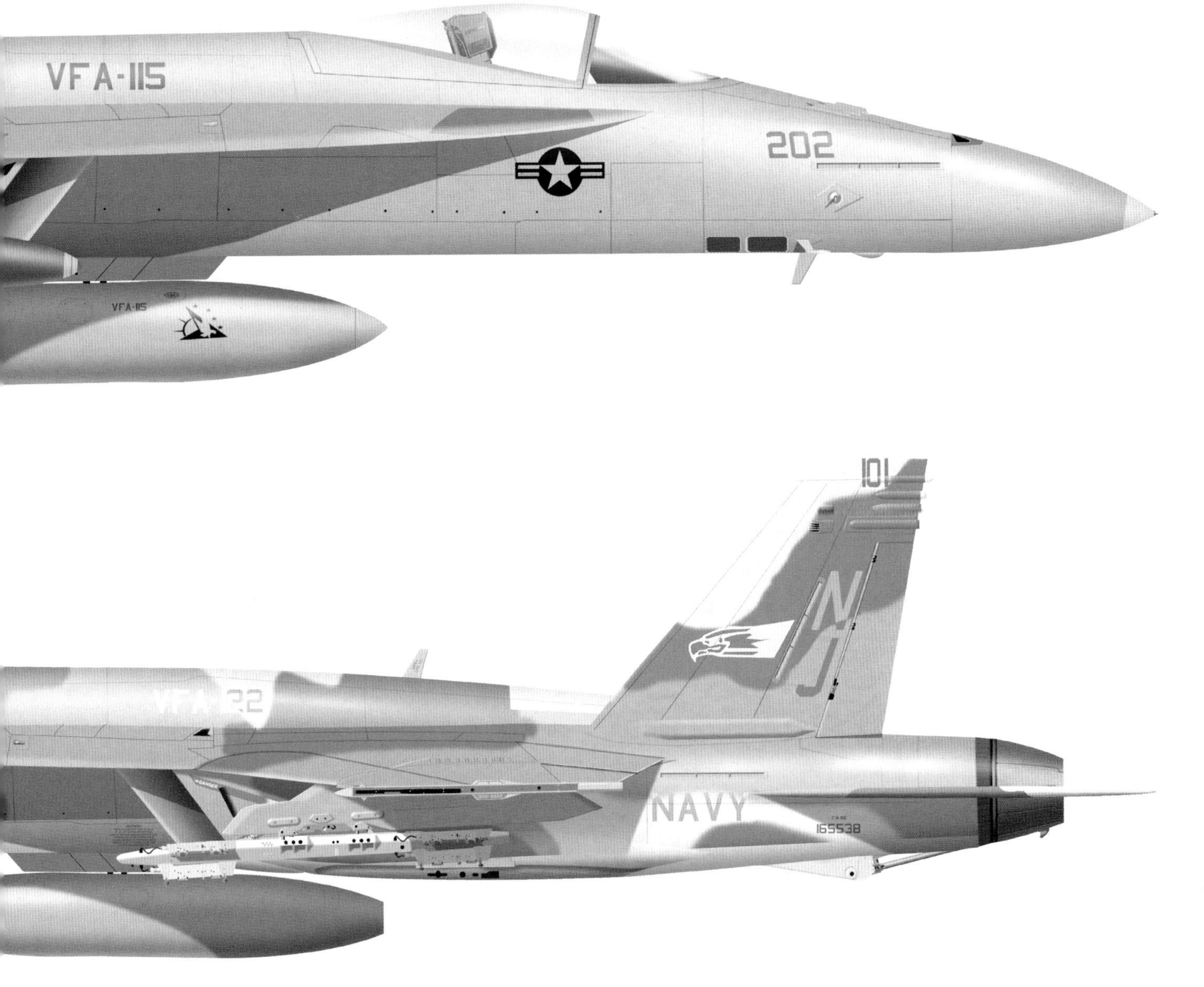

Boeing F/A-18E/F Super Hornet

of what was dubbed the 'Hornet 2' in June 1992 and the AX programme – which had since become the A/F-X – was cancelled in September 1993 (being superseded by JAST – see next chapter).

The Hornet 2 was given the in-service designations F/A-18E and F/A-18F for the single and two-seater variants respectively and its official name became Super Hornet (though to avoid confusion with the existing Hornet it was usually known as the 'Rhino' in service). A great many hopes rested on it too; it would need to replace the F-14 on fleet defence, the F/A-18C on air superiority, strike and reconnaissance, the S-3B as a tanker and the EA-6B on electronic warfare.

The first prototype or 'engineering and manufacturing development' aircraft made its flight debut on November 29, 1995, with testing commencing the following year and the type's first carrier landing taking place in 1997. Initial production began in March 1997 and full production began six months later.

That same year, following Boeing's acquisition of Rockwell's North American division, McDonnell Douglas merged with Boeing, with the latter being the surviving

BOEING F/A-18 SUPER HORNET

Boeing F/A-18E Super Hornet, AJ-302, VFA-87, USS *George H. W. Bush* (CVN-77), 2017. On June 18, 2017, Lt Cmdr Michael Tremel was credited with the first US air-to-air kill since 1999, after shooting down a Syrian Sukhoi Su-22.

company. As a result, the McDonnell Douglas F/A-18E/F became the Boeing F/A-18E/F.

The Super Hornet, as it turned out, was almost a completely new aircraft. The forward fuselage was the only part of the design carried over largely unaltered with the rest of the airframe, though resembling that of its predecessor, being entirely bespoke. However, structurally, it had 42% fewer parts than the original Hornet.

Its wings were 25% larger as well as being thicker to prevent flexing issues which had affected the original; sweepback was increased and the LERX were altered to account for the effect of the other changes on the type's manoeuvrability. The fuselage was elongated by 34in to accommodate more fuel, addressing concerns about the original Hornet's range, and extra space was allocated to the advanced avionics that the aircraft would need to carry.

These changes added a significant amount of weight but General Electric stepped up to provide more power in the form of the new F414-GE-400 turbofan engine – which generated 13,000lb-ft of dry thrust and 22,000lb-ft ▶

BOEING F/A-18 SUPER HORNET

Boeing F/A-18E Super Hornet, NF-200, VFA-27, USS *George Washington* (CVN-73), 2010. VFA-27 transitioned from the Hornet into the Super Hornet in 2004; the CAG bird of Carrier Wing 5 displays a fitting colour scheme.

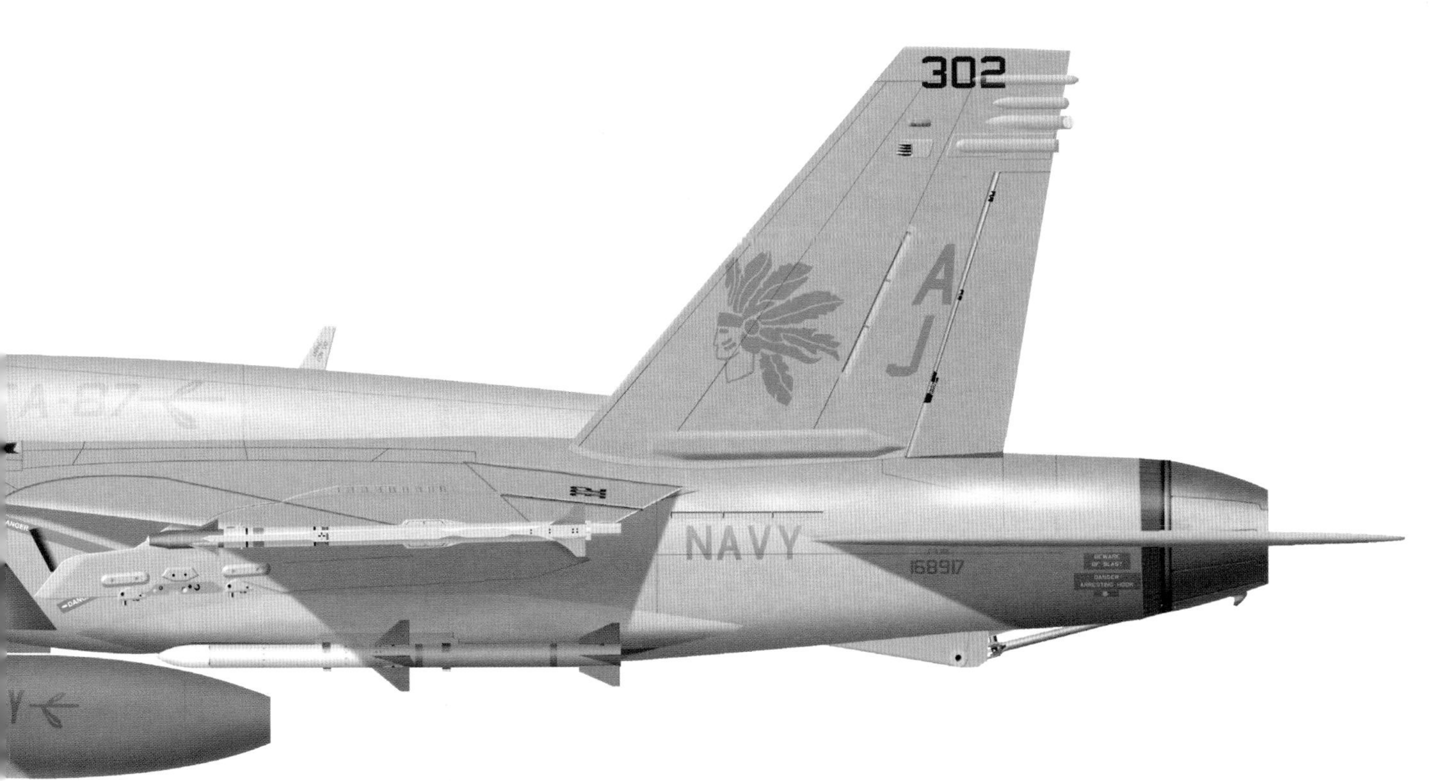

Boeing F/A-18E/F Super Hornet

BOEING F/A-18F SUPER HORNET

Boeing F/A-18F Super Hornet, NF-101, VFA-102, USS *Kitty Hawk* (CV-63), 2008. VFA-102 was equipped with the F/A-18F in 2002; this aircraft is configured to carry out buddy-buddy refuelling tasks.

with afterburner, compared to the original Hornet's F404-GE-402 which managed 11,000lb-ft dry and 17,750lb-ft with afterburner.

Once of the most distinctive visual differences between the Hornet and the Super Hornet is the latter's engine intakes. Where the former's were D-shaped, the latter's have a more conventional rectangular form. They also feature a baffle intended to reduce the aircraft's radar signature. The aerodynamic form of the F/A-18E and F was further tweaked to make it more difficult to detect and like the later Hornets it also made use of radar absorbent materials.

When the Super Hornet first appeared its avionics suite was essentially the same as the most recent upgrades to the F/A-18C, including the APG-73 radar – since this was already the latest equipment available, though further electronics improvements would be necessary before it could carry and launch all the new munitions intended for it.

The aircraft achieved initial operating capability in September 2001 with VFA-115 and on November 6, 2002, a pair of F/A-18Es were involved in a strike mission as part of Operation Southern Watch – monitoring and controlling

BOEING F/A-18E SUPER HORNET

Boeing F/A-18E Super Hornet, 1, Blue Angels, NAS Pensacola, Florida, 2020. It was announced in 2017 that the Blue Angels would be equipped with F/A-18 Super Hornets; the first aircraft was delivered to the team in July 2020.

southern Iraqi airspace. One of them dropped a 2000lb JDAM bomb on a bunker complex operating a pair of surface-to-air missile launchers.

F/A-18Es and Fs were initially unable to fulfil the reconnaissance role planned for them because the required Raytheon-developed digital system with infrared and electro-optical sensors, known as SHARP (SHAred Reconnaissance Pod), was delayed.

It finally made an appearance in prototype form during Operation Iraqi Freedom, the invasion of Iraq, in 2003. During the same conflict, the F/A-18E/F was able to demonstrate another of its capabilities to full effect – operating as a combat-ready tanker by carrying five external tanks.

In addition to the arrival of SHARP, the early Super Hornets received upgrades during the early 2000s which included updated cockpit displays and other electronics. From 2003, all F/A-18Es and Fs featured a completely new forward fuselage. Although it is visually near-identical to the original, it requires 40% fewer individual parts and provides better access to the aircraft's electronics for servicing.

Starting in 2005, new build F/A-18E/Fs were fitted with Raytheon's APG-79 Active Electronically Scanned ▶

Boeing F/A-18E/F Super Hornet

Array (AESA) radar as part of Block II upgrades. This system has a detection range three times greater than that of its predecessor and features a Synthetic Aperture Radar mode with a resolution five times greater than the APG-73's. It is able to detect much smaller objects, such as missiles, and can track aerial targets beyond the range of its own missiles.

Block II machines also carry better countermeasures in the form of a BAE Systems on-board jammer, a fibre-optic towed decoy, the AN/ALR-67 radar warning receiver and AN/ALQ-214(V)3 Integrated Defensive Electronic Counter-Measures system.

Most Block II machines are F/A-18Fs so that the back-seaters – known as weapon systems officers in the Super Hornet, rather than radar intercept officers – can track, assess and act upon the flood of information delivered by the aircraft's sensors and via its network connections.

The first Super Hornets with AESA installed were declared 'safe for flight' on October 27, 2006, and on May 18, 2007, the first aircraft to feature the Joint Helmet Mounted Cueing System was

BOEING F/A-18F SUPER HORNET

Boeing F/A-18F Super Hornet, SD-223, VX-23, NAS Patuxent River, Maryland, 2018. The main mission of VX-23 is testing fixed wing aircraft, including the Super Hornet.

delivered to VFA-213. This provides the aircraft's crew with enhanced situational awareness, effective target designation up to 80° either side of the aircraft's nose and integrated night vision goggles.

In 2015, the Super Hornet was upgraded to carry the AN/ASG-34 Infrared Search & Track (IRST) imager in front of its centreline fuel tank, which provides a passive sensing capability even if the aircraft's radar system is being jammed. One of its particular functions is to detect stealth aircraft through the use of thermal imaging rather than via radar returns.

The final Block II Super Hornets were delivered during the spring of 2020. A grand total of 608 Block I and Block II aircraft have been built to date – 322 F/A-18Es and 286 F/A-18Fs. At the time of writing, the Navy had 78 new Block III Super Hornets on order from Boeing – the key features of which will be the introduction of conformal fuel tanks, touchscreen displays in the cockpit and a reduced radar cross section. In addition, plans were in train to upgrade a significant number of the remaining 540 Super Hornets to Block III standard. Boeing announced the delivery of the ▶

BOEING F/A-18F SUPER HORNET

Boeing F/A-18F Super Hornet, AB-106, VFA-11, USS *Enterprise* (CVN-65), 2011. VFA-11 was equipped with the Super Hornet in 2005 and made several deployments with the new aircraft, including in support of Operation Enduring Freedom in 2011.

Boeing F/A-18E/F Super Hornet

first two Block III aircraft to the Navy for testing on June 17, 2020.

EA-18G SUPER GROWLER

With the Grumman EA-6B Prowler, the Navy's standoff electronics countermeasures aircraft, approaching retirement in 2000, preparations began to transfer this role to the Super Hornet. A firm commitment was made to an ECM version of two-seat F/A-18F in late 2002 as the EA-18G Growler. A heavily modified F/A-18F commenced flight testing as a development aircraft in November 2003 and the first prototype flew during the summer of 2006. The Navy took delivery of the first production examples during the spring of 2008 and initial operational capability was reached in September 2009.

The Growler was built as part of the Block II series but with the Improved Capability (ICAP) III system developed for the Prowler as the starting point for its countermeasures package. It features the modernised AN/ALQ-218(V)2 tactical jamming receiver (replacing the 20mm cannon in the nose) and other Airborne Electronic Attack subsystems, including a total

BOEING F/A-18F SUPER HORNET

Boeing F/A-18F Super Hornet, AG-200, VFA-103, USS *Harry S. Truman* (CVN-75), 2016. The Super Hornets of VFA-103 participated in Operation Inherent Resolve in 2016, racking up an impressive mission record as displayed on the nose of this aircraft.

BOEING F/A-18F SUPER HORNET

Boeing F/A-18F Super Hornet, Advanced Super Hornet Demonstrator, 2013. The new Block III (Advanced) Super Hornet features several improvements, including AESA radar and other avionics and systems enhancements. Here the demonstrator exhibits two of the more externally visible improvements: conformal fuel tanks (CFT) and Enclosed Weapons Pod.

of 66 antennas, making it 1400lb heavier than the standard F/A-18F. The antennas are most visible on the dorsal deck, aft fuselage and nose barrel, with wingtip pods for the AN/ALQ-218(V)2 in place of the air-to-air missile rails.

The Growler also has a Raytheon AN/ALQ-227 Communications Countermeasures Set to locate and jam enemy communications – consolidating what had been four 'boxes' on the Prowler into just one. Another innovation is ITT Electronic Systems' Interference Cancellation System to enhance the crew's situational awareness.

On the underwing pylons, the EA-18G can carry up to five ITT ALQ-99 transmitter pods to cover an array of different jamming bands. One of the greatest challenges when designing the Growler's systems was the reduction in crew from the Prowler's three or four to just two. Numerous workload studies were carried out and many systems which previously required manual intervention were automated. A total of 160 Growlers were bought, with the last one being delivered during the summer of 2020.●

LOCKHEED MARTIN F-35C LIGHTNING II

The carrier variant of the controversial Lightning II stealth strike fighter, the F-35C, is the largest and heaviest of the three versions currently in production. Its electronics are highly advanced but questions have been raised about its performance in flight and its abilities have yet to be tested in a real-world combat situation.

The F-35 is the end result of a multitude of development programmes stretching back to the late 1980s – particularly Common Affordable Lightweight Fighter (CALF) and Joint Advanced Strike Technology (JAST).

CALF was begun in 1993 with the intention of creating a VSTOL aircraft for the USMC and British Royal Navy – while also providing a low-cost, low-maintenance fighter for the USAF. At the same time, the US Department of Defense was conducting a bottom-up review of ongoing programmes which resulted in the cancellation of the Navy's A/F-X, as mentioned previously.

The Super Hornet would still be built – but now as a stopgap for a new multirole strike aircraft to be designed and built under JAST, rather than A/F-X. JAST then absorbed CALF. Procurement of the Navy's F/A-18C/D and the USAF's F-16 was reduced in anticipation of both types' ultimate replacement by the new JAST fighter.

Design studies from manufacturers McDonnell Douglas, Northrop Grumman, Lockheed Martin and Boeing were submitted for JAST in 1993 and the programme was renamed Joint Strike Fighter in 1994. The goal was to create an affordable strike aircraft that would also be second only to the F-22 in the air supremacy role. In order to meet the requirements of such a diverse range of services and roles it would need to be available in three forms – one that took off and landed conventionally, one capable of short take off and vertical landing and a CATOBAR (carrier-based catapult assisted take-off but arrested recovery) version primarily for the US Navy. In November 1995 the UK agreed to pay $200 million to buy in as a partner on the project.

A year later, on November 16, 1996, both Lockheed Martin and Boeing received $750 million contracts to develop prototypes. Lockheed Martin's design was given the designation X-35 and Boeing's X-32. Two X-35 prototypes were developed – the X-35A, later converted into the X-35B, and the X-35C, which had larger wings. The X-35A completed its first flight on October 24, 2000, and the process of converting it into the X-35B commenced on November 22, 2000. The X-35C made its first flight on December 16, 2000. During final qualifying Joint Strike Fighter flight trials, the X-35B STOVL (short take-off, vertical landing) aircraft was able to take off in less than 500ft, go supersonic, then land vertically – which Boeing's equivalent design was unable to match.

Lockheed Martin was declared the winner and awarded a contract for system development and demonstration on October 26, 2001. The JSF programme was by now being jointly funded by the US, UK, Italy, Holland,

LOCKHEED MARTIN F-35C LIGHTNING II

Lockheed Martin F-35C Lightning II, CF-01, 100, VX-23, NAS Patuxent River, Maryland, 2011. This was the first F-35C for the US Navy.

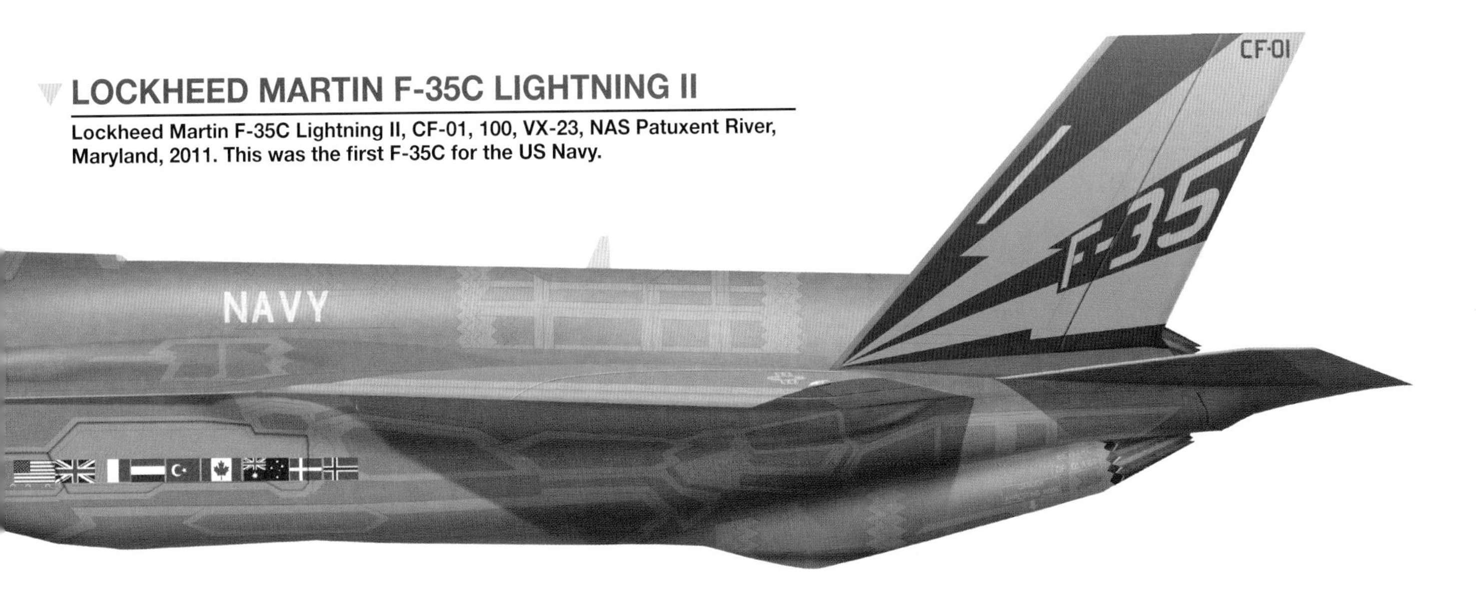

LOCKHEED MARTIN F-35C LIGHTNING II

Lockheed Martin F-35C Lightning II, CF-05, SD-75, VX-23, USS *Nimitz* (CVN-68), 2014. As with other fighters of the US Navy, the F-35C is equipped with a retractable in-flight refuelling probe.

Canada, Turkey, Australia, Norway and Denmark.

During further development, Lockheed Martin slightly enlarged the X-35, stretching the forward fuselage by 5in to make additional space available for the avionics. The tailplanes were correspondingly moved 2in further back to retain balance. The upper fuselage was raised by an inch along the centre line too. Parts manufacture for the first prototype began on November 10, 2003.

The X-35 had lacked a weapons bay and adding one resulted in design changes which increased the aircraft's weight by 2200lb. Lockheed Martin addressed this by increasing engine power, thinning the airframe members, reducing the size of the weapons bay itself and the size of the aircraft's fins. The electrical system also underwent changes, as did the section of the aircraft immediately behind the cockpit. All this succeeded in reducing weight by 2700lb but at a cost of $6.2 billion and 18 months of additional development time.

The Navy's F-35C is powered by a single 50,000lb-ft thrust Pratt & Whitney F135-PW-100 engine which, while it lacks a supercruise function, does enable the aircraft to fly at Mach 1.2 for 150 miles without using its afterburner. With afterburner, it has a top speed of Mach 1.6.

Its dimensions are somewhat larger than those of the F-35A and B, measuring 51.5ft long and 14.7ft high with a wingspan of 43ft and wing area of 668ft². In addition, its wingtips can be folded to save spare on board a carrier. Internal fuel capacity is 19,750lb – providing the F-35C with greater range than either of its siblings and its overall empty weight is 34,800lb, compared to 32,300lb for the F-35B and 29,300lb for the F-35A. Maximum weight is 70,000lb and maximum g-rating is 7.5 g, compared to 7 g for the F-35B and 9 g for the F-35A.

Lacking the internal 25mm GAU-22/A cannon of the F-35A, the F-35C's ▶

LOCKHEED MARTIN F-35C LIGHTNING II

Lockheed Martin F-35C Lightning II, CF-02, VX-23, NAS Patuxent River, Maryland, 2018. When stealth is not essential, F-35Cs can carry loads externally as well as internally.

LOCKHEED MARTIN F-35C LIGHTNING II

Lockheed-Martin F-35C Lightning II, NJ-101, VFA-101, AFB Eglin, Florida, 2012. VFA-101 became the Fleet Replacement Squadron for the F-35C; it received its first aircraft in 2012 and operates from Air Force Base Eglin.

LOCKHEED MARTIN F-35C LIGHTNING II

Lockheed Martin F-35C Lightning II NE-411, VFA-147, NAS Lemoore, California, 2019. In early 2019, two F-35C pilots, one each from VFA-125 and VFA-147, became the first to graduate from the Navy Strike Fighter Tactics Instructor course, known as Topgun in popular culture.

CF-02
NAVY
101
NJ
GRIM REAPERS
VFA-101
NAVY
F-35C
168733
411
NE
NAVY
F-35C
169600

Lockheed Martin F-35C Lightning II

LOCKHEED MARTIN F-35C LIGHTNING II

Lockheed Martin F-35C Lightning II, 200, NAWDC, NAS Fallon, Nevada, 2020. This aircraft from the Naval Aviation Warfighting Development Center shows the weapons bays which allow for stealth missions.

LOCKHEED MARTIN F-35C LIGHTNING II

Lockheed Martin F-35C Lightning II, CF-03, SD-73, VX-23, USS *Nimitz*, 2014. Aircraft from Air Test and Evaluation Squadron 23 made the first arrested carrier landing with the F-35C.

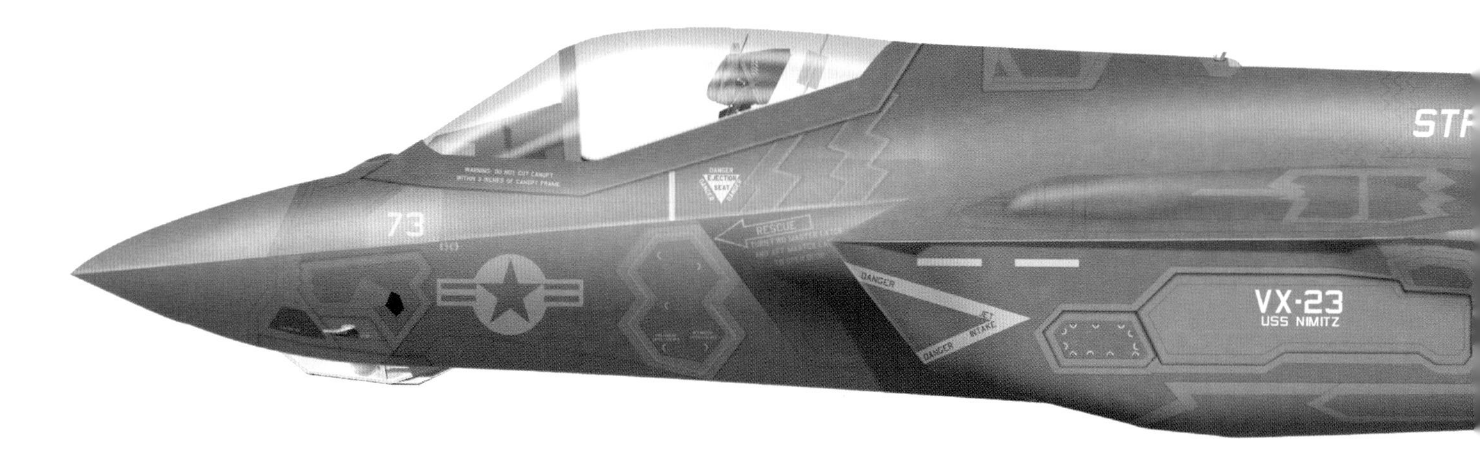

standard internal weapons load is two AIM-120C/D air-to-air missiles plus two 2000lb GBU-31 JDAM guided bombs. However, the cannon can be carried externally as a pod with 220 rounds if necessary. Four underwing pylons can carry AIM-120s, AGM-158 cruise missiles and guided bombs. Two further near-wingtip pylons are designed for the AIM-9X sidewinder and AIM-132 ASRAAM. Using both internal and external stations an air-to-air missile load of eight AIM-120s and two AIM-9s is possible.

Inside its cockpit, the F-35 has a 20x8in touchscreen, a speech-recognition system, a helmet-mounted display, a right-hand side stick controller, a Martin-Baker ejection seat and an oxygen generation system derived from that of the F-22. Due to the helmet display, the aircraft does not have a HUD. Its radar is the AN/APG-81 developed by Northrop Grumman Electronic Systems with the addition of the nose-mounted Electro-Optical Targeting System. The F-35's electronic warfare suite is the AN/ASQ-239 (Barracuda) with sensor fusion of radio frequency and infrared tracking, advanced radar warning receiver including geolocation of targeting of threats, and multispectral image countermeasures.

The aircraft has 10 radio frequency antennas embedded in its wings and

tail. Six passive infrared sensors are distributed across the F-35 as part of Northrop Grumman's AN/AAQ-37 distributed aperture system. This provides missile warning, reports missile launch locations, detects and tracks approaching aircraft and replaces traditional night vision devices.

The first F-35, AA-1, was rolled out on February 20, 2006, and the type was formally given the name Lightning II on July 7, 2006 – the 'II' making it a spiritual successor to both the Lockheed P-38 Lightning of the USAAF during the Second World War and the British English Electric Lightning of the Cold War. AA-1 made its flight debut on December 15, 2006, and while A and B variants flew sooner, the F-35C made its flight debut on June 6, 2010, with the Navy receiving its first examples in 2011. Deliveries to the first US Navy F-35C squadron, VFA 101, took place in 2013. Initial Operational Capability was declared on February 28, 2019, and the first F-35C pilots to complete the famous Topgun training course – from VFA 125 'Rough Raiders' and VFA-147 'Argonauts' – did so in June 2020. The Navy currently has 273 F-35Cs on order.

The F-35 programme has proven highly controversial over the years – with cost overruns, delays and accusations ▶

of industrial espionage. The aircraft's performance has also come under intense scrutiny, with some detractors assessing its manoeuvrability as mediocre and its weapons payload as inadequate. Its ability to counter the threat posed by increasingly sophisticated foreign types such as the Russian Sukhoi Su-57 and Chinese J20 has also been doubted.

It was reported in 2010 that the F-35 took 43 seconds longer to accelerate from Mach 0.8 to Mach 1.2 than the F-16 it is intended to replace. Concerns have also been raised about poor visibility from the F-35's cockpit, gaps in its stealth coating, a fire risk arising from fuel tank vulnerability, engine problems, a high false alarm rate with the F-35's helmet, software failures and deficiencies, and low availability and reliability.

In July 2015, a leaked Lockheed Martin report showed that the F-35 was less manoeuvrable than an older F-16D fitted with wing tanks. The Pentagon responded that the F-35 was not intended for dogfighting but would instead disrupt enemy advanced air defences. It also has sensors capable of detecting and engaging enemy aircraft at beyond-visual-range, negating the need for close-in dogfighting. It has been noted, too, that the F-35 may be able to offer greater manoeuvrability in the future.

That said, when F-35s took part in their first large-scale aerial combat training exercise – Red Flag 17-1 – during the spring of 2017, against Dutch F-16s, they achieved an impressive kill ratio of 20-1. USMC F-35Bs apparently managed 24-0 during one exercise. All this was despite the fact that the F-35s were actually flying simulated strike rather than air superiority missions.

With the F-35C only just beginning its career as a front line combat aircraft, it remains to be seen whether history will come to regard it as a high-tech success story or a costly white elephant.●

LOCKHEED MARTIN F-35C LIGHTNING II

Lockheed Martin F-35C Lightning II, NJ-412, VFA-125, USS *Carl Vinson* (CVN-70), 2018. F-35Cs from VFA-101 and VFA-125 operated from the USS *Carl Vinson* in December 2018, during fleet replacement squadron carrier qualifications.

LOCKHEED MARTIN F-35C LIGHTNING II

Lockheed Martin F-35C Lightning II, XE-107, VX-9, Edwards AFB, California, 2017. VX-9 squadron operates a detachment of F-35Cs at Edwards Air Force Base.

LOCKHEED MARTIN F-35C LIGHTNING II

Lockheed Martin F-35C Lightning II, VFC-13, NAS Fallon, Nevada, 2028. With the emergence of stealth fighters originating from other countries it's plausible to assume that the US Navy will have adversary units operating the F-35C; here is a speculative scheme for such an aircraft.

412
NJ
VFA-125
NAVY
F-35C
169303
107
XE
VX-9
NAVY
F-35C
169424
VFC-13
NAVY

Great reads from £7.99!

132-page full colour, perfect bound bookazines

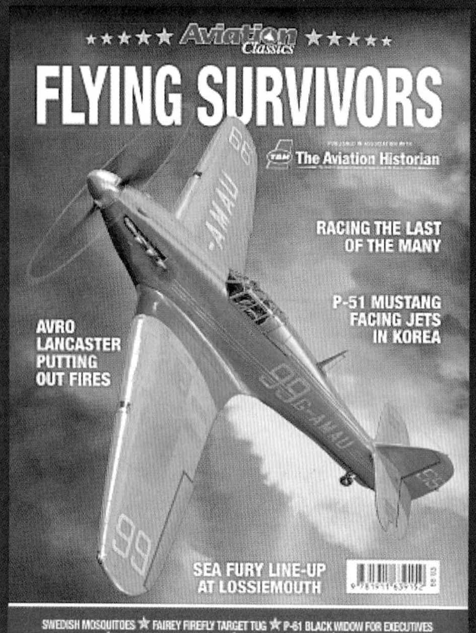

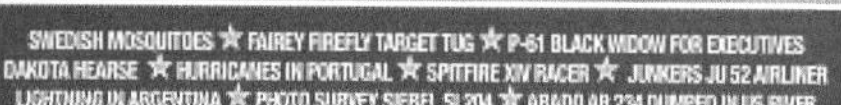

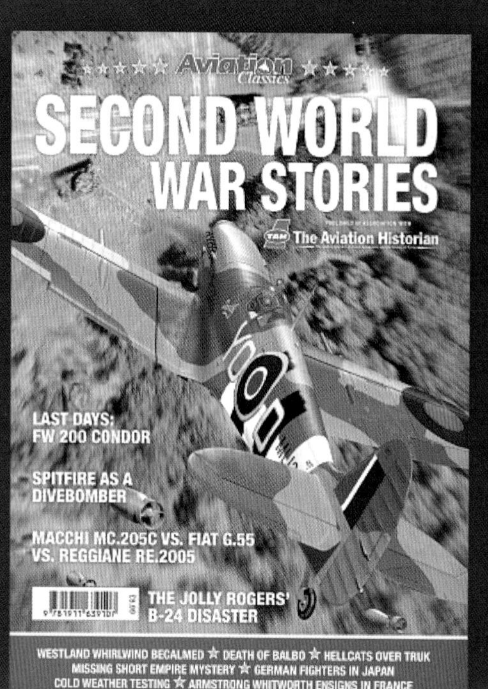

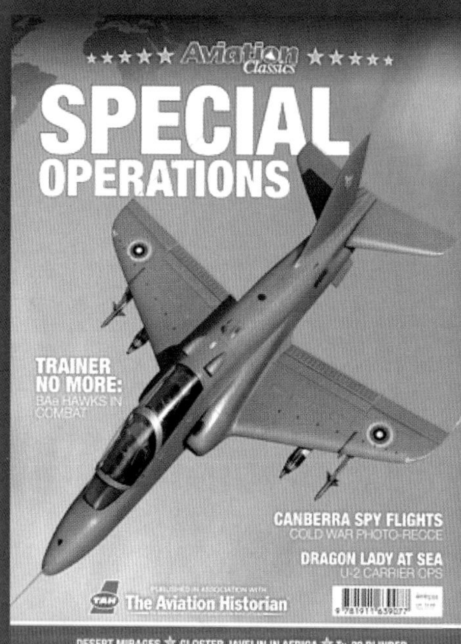

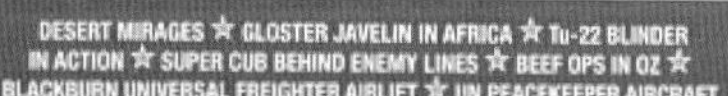

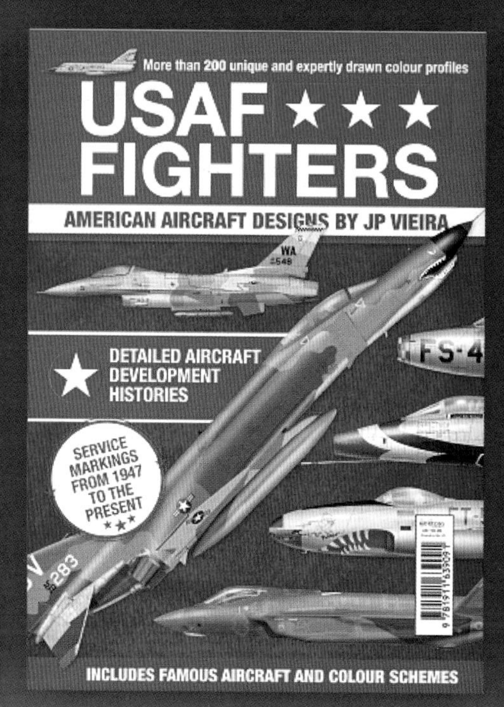

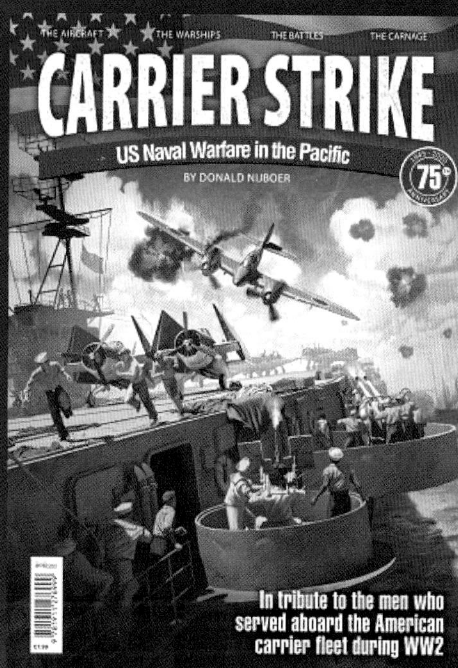

ORDER TODAY FROM CLASSIC MAGAZINES:

www.classicmagazines.co.uk/thebookshelf

CALL: 01507 529529

classic magazines

Also available at WHSmith